NO
COUNTRY
FOR A
WOMAN

Dolly's bookplate. (Author's photo, from a copy
of *The Country of the Orinoco* that must have
belonged to her)

NO COUNTRY FOR A WOMAN

THE ADVENTUROUS LIFE OF
LADY DOROTHY MILLS
EXPLORER AND WRITER

JANE DISMORE

The History Press

First published 2025

The History Press
97 St George's Place, Cheltenham,
Gloucestershire, GL50 3QB
www.thehistorypress.co.uk

British Library Cataloguing in Publication Data.
A catalogue record for this book is available from the British Library.

ISBN 978 1 80399 614 1

Typesetting and origination by The History Press
Printed and bound in Great Britain by TJ Books Limited, Padstow, Cornwall

MIX
Paper | Supporting
responsible forestry
FSC® C013056

Proudly supporting

Trees for Life

CONTENTS

INTRODUCTION

When Lady Dorothy Mills was a young girl, a female relative told her she would never be beautiful, so she had better be interesting – and as a bold, determined woman from one of Britain's oldest political and literary families, the Walpoles, she became just that. Rather than following the path expected of her class and sex, she took control of her destiny to become the most popular female explorer of the time. With a deep interest in people and an insatiable curiosity about the world, she turned her experiences into highly acclaimed travel books and escapist novels and in the 1920s and 1930s was seldom out of the public eye; newspaper readers also enjoyed her feminist features at a time when women were finding new freedoms in the period between the two world wars.

How extraordinary, then, that this is the first book about her life. Today, her name seldom comes to mind when women travellers are discussed, even though she journeyed alone (and separately from her husband) during the volatile period following the Great War, to countries where European women had rarely, if ever, ventured. Sometimes, she was the first white woman to do so, enduring physical and mental challenges and life-threatening hazards. A woman of her time in certain ways, ahead of it in others, her life provides a prism through which we can see and understand better the issues of the day – race, gender, colonialism, class – while also reminding us of what draws us together rather than what divides us.

So, why is she not better known? She has been referred to, often sketchily, even inaccurately, in compendiums. Her books were relied on during her lifetime as a source of information and, more recently, for academic study, but now are out of print, her passionate prose and self-deprecating humour accessible only to those who know of their existence and are willing to hunt them down. She has devoted fans on the internet who know a little about her life but cannot find more. It is as though something has conspired to keep her out of history. Ultimately, though, the memory of a person can be kept alight only by the living, and if no one does that, there is only darkness.

Born into the patriarchal society of the late Victorian era, she determined not to be limited by it or overlooked. 'I must *be* something,' she wrote. She certainly was, and I wanted to make her, like the woman in her novel *Phoenix*, rise again.

It has not been an easy task. That she was not publicity-shy and was prolific in her writing have been gifts in helping to piece together a life in which she underwent a metamorphosis from a 'useless young creature' to an inspiration for twentieth-century women, even while affected by personal tribulations. To enable me to write this book, I am deeply grateful to the historian and biographer Antonia Fraser, whose interest in my subject led to my receiving her eponymous award (in the form of an Authors' Foundation Grant), administered by the Society of Authors. Lady Dorothy Mills can live once more.

CAST INTO THE DARKNESS

By June 1916, the grimness of the Great War was changing the face of Britain's cities. Since the first air raids, a year earlier, London's Underground stations were becoming unofficial shelters for thousands and locals were wearily becoming accustomed to the dreary inevitability of blackouts. Conscription was newly in force, prompting angry demonstrations in Trafalgar Square. But every now and then a reminder of life as it was and might be again, of hope triumphing over misery, would appear above the parapet and give people a reason to smile.

On 22 June, a vision in a gold and white wedding gown paused outside St Paul's Church, Knightsbridge, to let the sunlight catch the gold leaves of her headdress and smile a greeting to reporters, before joining the waiting groom. Lady Dorothy Rachel Melissa Walpole, aged 27, petite and bold, wanted to make a statement to those naysayers who held little hope for her happiness. She proudly claimed to be the first London bride to wear a gold wedding dress, her decision typical of her originality, even though the cost meant it would be her only evening frock for the foreseeable future other than those she made herself.

If Dolly, as she was known, was a bright display, her fiancé was a more sober reminder of the times. Captain Arthur Frederick Hobart Mills, nearly 29, wore the uniform of his regiment, the Duke of Cornwall's Light Infantry. At 6ft 2in tall, with dark brown hair and heavy-lidded brown eyes, he was a handsome figure and part of the British Expeditionary Force in France until he was shot in his right ankle, an incident he described in his recently published book, *With My Regiment*,[1] and which had seen him invalided back to England.

Not only was the wedding dress unusual, but the ceremony was too, in which tradition played little part. Their engagement had been announced just a couple of weeks earlier and no wedding invitations were issued; instead, *The Times* announced simply that all their friends were welcome at the church. Unusually, the bride had no bridesmaids, a decision that was unlikely to offend anyone. When Dolly was 4, her only sibling, Horatio, had died at the age of 2 and she 'tactlessly remained the only child'.[2]

The groom's best man was a cousin and fellow officer, and on that side too, there was no one who could reasonably be upset by not having a role to play. His parents' only child, Arthur was 2 when his mother died; his father, the Reverend Barton Reginald Vaughan Mills, had remarried and now Arthur had three half-siblings, George, Agnes and Violet, all of whom were present.[3]

The ring that Arthur placed on Dolly's finger was fashioned not from traditional gold but from lead, taken from the bullet that had wounded him. After the ceremony, there was no formal reception, although that was not unusual for wartime. Apart from Dolly's dress, the simplicity of the occasion was commended by the newspapers as being a splendid example of frugality in the face of an uncertain wartime economy.

While the event was covered by much of the press, a further break with tradition was glossed over, if it was mentioned at all – the fact that the bride was given away, not by her father but by Arthur's uncle.[4] It was not that her father was dead, far from it, but Robert Horace Walpole, 5th Earl of Orford, did not approve of his daughter's choice of spouse, and Dolly could not look to her mother, Louise, for support, for she had died too soon.

The real reason for the simplicity of the wedding had less to do with sensible wartime husbandry than the fact that the earl would not pay for it. Worse than that, he had disinherited Dolly. The groom's lack of a title had likely not impressed him, yet Arthur was from an aristocratic family. His late mother, Lady Catherine Mary Valentia Hobart Hampden, was a sister of the 7th Earl of Buckinghamshire and Arthur's great-grandfather had been the 6th Earl. In the county of Norfolk, where the Walpoles and the Hobart Hampdens had owned estates for generations, the families were practically neighbours. From the age of 13, Arthur had been educated at Wellington College, a highly respected and forward-looking private school in Berkshire. If it was of any relevance to Dolly's future happiness, his father was the grandson of a baronet and his stepmother the daughter of a Scottish peer.

But for the Earl of Orford, whose ancestors included Britain's first prime minister, Sir Robert Walpole, and his son, the literary giant Horace Walpole, known as much for his eccentric house, Strawberry Hill, as for the first gothic novel, *The Castle of Otranto*, it seems that Captain Mills, in a family to which Admiral Lord Horatio Nelson had also belonged, had insufficient status or prospects. Certainly, Arthur's financial position played a major part in the earl's disapproval. As Dolly explained:

> I fell in love, with a young man possessing most of the world's assets except money. But that 'Except' had a capital 'E.' It was the one unforgivable sin, and was visited with everything old-fashioned and unpleasant that nothing but the Inquisition or an old-fashioned family could have devised. Marriage or disinheritance, that was the choice that lay before me, exacerbated by the advent of the Great War.[5]

Three years of 'family warfare' had preceded her wedding, ending in what she called a draw: on the one hand, she had done what she had intended, on the other, she was 'definitely cast into utter darkness, to become the Outlier that I have ever since remained'. Her marriage was a brave move on her part: 'I had no trousseau, we had no prospects and no money, scarcely enough even to pay for the wedding celebrations.'[6] Her first novel, *Card Houses*, was to be published a few months after the wedding,[7] but there was no telling yet how well it might sell.

Dolly had ignored the advice of those who recommended a registry office, choosing instead a church wedding, 'and a fashionable one too', so that if it came to it, it would be her 'last defiance to a sceptical world'. Having never arranged a wedding before, and with no one to help her, she was desperately tired on the day but considered it well worth the trouble and money it had cost, 'for it proved even an Outlier has friends and well wishers angelic in their kindness and good will'.[8]

Indeed, the earl's attitude had not deterred the attendance of around 100 family members and friends, mostly from the upper echelons of society. Arthur's family was well represented, with his other uncle, the Earl of Buckinghamshire, heading the maternal side, while a good number of Walpole relatives supported Dolly. Among them were Ralph and Meresia Nevill, two of the children of her late great-aunt Lady Dorothy Fanny Nevill. Another colourful and clever Walpole, she had survived the scandal of being caught in 1847 with a notorious rake who refused to marry her, hurriedly marrying instead her much older but kindly cousin, Reginald Nevill, and becoming a multi-talented and well-known figure in Victorian society, even though the Queen herself banished her from court.

The creatives and eccentrics on the Walpole side of the family gave Dolly an interesting heritage and a wealth of anecdotes, which made her an amusing raconteur. Both her parents were avid travellers, taking their daughter with them as she grew older and showing her a fascinating world which expanded her already curious mind. Her French-educated mother, Louise Melissa Corbin, was the daughter of American mining and railroad magnate Daniel Chase Corbin, whom Dolly visited as a child and came to love the USA as her second home. 'The great expanse of the Far West, the grizzly bear in the Yellowstone Park [...] the war dances of Red Indians,' she recalled, 'the Mormons of Salt Lake City [...] stories of hold-ups in the newspapers [...] all these added fuel to the youthful fire of adventure.'[9]

This love of travel, combined with her literary ability, would prove to be a lifeline. In their first year of marriage, she had what she called her 'first taste of the economic problem'. Despite her burgeoning talent, including the publication of her first poem when she was in her teens, she had been expected to marry well, as were most young women of her background, without the need to worry about 'petty household and personal economies and makeshifts', and she was raised for that destiny. She had never learned how to do her hair without her maid or how to mend holes in her stockings. Her first attempt to lace up her own boots gave her 'a headache and an intense desire to cry'. 'In fact,' she admitted, 'never had there been such a useless young creature, till necessity turned me into a very fair Jack-of-all-trades.'[10]

When Arthur rejoined the war to serve in the Palestinian campaign, Dolly faced grim months of financial worry and 'grinding anxiety in a world where nothing seemed stable, where the future did not bear dwelling on'. She soon acknowledged that she did not have it in her 'to remain placidly in the background of life as "just" the wife of a poor man, however charming or clever he might be'. She also became increasingly aware that her 'purely decorative upbringing' had left her ill equipped to contribute usefully, and she saw other young women doing heroic things at home and in France. 'Something hammered insistently in my head that I must do something,' she wrote, 'have some money, live life to the full, must *be* something.'[11]

Continuing family discord exacerbated her anxiety. In September 1917, her 63-year-old father, determined to secure a male heir, took a second wife. The pretty daughter of a country vicar, at 25 Emily Gladys Oakes was three years younger than her stepdaughter. It would not be long before Dolly came to a realisation: 'I was not to be reinstated among my own kind […] I was an Outlier from my tribe.'[12] Little wonder that she determined to make a mark on her own terms in a predominantly male world that had changed little since she was born.

ONLY A GIRL

'The second little disappointment tonight.'[1] The doctor who uttered these words as he delivered Dolly on 11 March 1889 was fully aware of the expectations of aristocratic clients like her father. No doubt Robert Walpole congratulated his wife on the safe delivery of their first child, born ten months after their marriage, but really, it was a boy he wanted. Walpole himself was heir to his uncle, the 4th Earl of Orford, by virtue of the earl's lack of a legitimate son. While his wife had given him two very capable daughters, his only son was the result of his adulterous relationship with the married Countess of Lincoln (a highly scandalous and ultimately unhappy affair), and therefore could not inherit.

England's class structure and the customs concerning aristocratic inheritance were a concept that was alien to Dolly's American mother, Louise, but one with which, on meeting Walpole, she had necessarily become acquainted. Not that everything in her new life was unfamiliar to her. Louise was the daughter of the immensely wealthy Daniel Chase Corbin who, after success in mining and banking, had recently joined his brother in the railroad business, building his first railway at a time of major expansion in the USA. As such, she was no stranger to the advantages that money conferred or the divisive social codes created by those who considered themselves the cream of American society, however meritorious they liked to think it was.

When Louise was 9, Corbin sent her and her younger sister, Mary, to live in Europe with their mother, Louisa; while his daughters' horizons

15

would be expanded, the main reason was his wife's poor health. Louise was educated in Germany, then France, where Mrs Corbin took an apartment on the Champs-Élysées in Paris and embraced the culture. There, Louise took her exams and showed particular proficiency in French, German and music; maths was more arduous, and saw her having to get up at dawn to keep abreast of the work.

The contents of a daybook Louise owned before her marriage give a glimpse of the happy sociability of her young life. Corbin started the book off for her in April 1885, when she was 18, with a note of fatherly advice: 'Be polite to your equals, kind to your inferiors and always honest and true.'[2]

Tributes to Louise included a love letter in French, *'un souvenir d'amour'* from a smitten Georges Roussel in August 1886. Exquisite miniature artworks – watercolours and pen and ink drawings – by friends who were not professionals but who had learned enough of the genteel art to leave their affectionate mark, depict a shared place, a fond memory, a joke. A sketch by a female friend shows a young woman at sea in a boat whose sails are giant butterflies controlled by a handsome swain. A charming ink drawing by Louis L. Odero, dated September 1885, depicts a group of young men and women enjoying outdoor pursuits among the mountains of Glion, Switzerland, playing tennis and croquet or linking hands as they help each other up a slope, overlooked by the elegant Victoria Hotel. A British Cabinet minister's painting of a country cottage in September 1887 is accompanied by a letter promising Louise he will complete it if, on his return, he finds her still at Inverlochy Castle in the Scottish Highlands, where they were guests of Lord and Lady Abinger.

An amusing sketch painted that month, also of Scotland, shows two women on horseback confidently riding side-saddle up Ben Nevis, along a narrow ridge with a precarious drop, while a man hangs on to the tail of the second horse and is pulled along; Louise was a devotee of riding and most outdoor activities, a love she would pass on to her daughter. On the lochs, she honed her skills in angling, reflected in a little watercolour by 'CB', probably Clifford Bingham, a popular poet whose offering 'For Want of You' also graces her book. But any young man who dreamed of winning the heart of the pretty, talented – and very wealthy – Miss Corbin would have his hopes dashed when in January

1888, aged 21, she became engaged to Robert Horace Walpole, twelve years her senior.

They had met just four months earlier at that Inverlochy house party, at which their hostess was delighted to pair them. Like Louise, Lady Abinger was American, the daughter of a commodore in the US Navy. As Helen Magruder, she had met Lord Abinger in Canada, where he was stationed with his Scottish regiment, and married him in 1863, becoming one of the first Americans to marry into the British peerage; her arrival in society was truly confirmed when Queen Victoria stayed at their castle ten years later. Happy in her position, she saw the opportunity to help a fellow countrywoman find hers.

Walpole was not a tall or particularly distinguished-looking man, but he must have presented a dashing persona to Louise. Like his late father, Commander Hon. Frederick Walpole, and their ancestor, Admiral Lord Horatio Nelson, he had enjoyed an eventful career in the Royal Navy, beginning at the age of 14. At a time when it was conducting its crusade against the slave trade, his adventures included being shipwrecked for three months in 1871 on St Paul's Island in the Indian Ocean and surviving on gulls' eggs and seaweed. During the Russo-Turkish War he was aide-de-camp to Suleiman Pasha and, on leaving the Royal Navy as a lieutenant, aged 24, he was attached to the Earl of Rosslyn's Special Embassy to attend the marriage of King Alfonso XII of Spain. That was followed by his appointment as private secretary and later attaché to his relation, Sir Henry Drummond Wolff, accompanying him to Egypt in 1886 when Wolff was High Commissioner. As heir to the earldom of Orford, his prospects were promising and the title historical, having been created for Sir Robert Walpole, Britain's first prime minister in the early eighteenth century. He was also a keen traveller and proficient angler, both of which he had in common with Louise.

They married in May 1888 at the British Embassy Chapel in Paris. Five years later, Louise would find herself (along with her friend, Lady Abinger) in a list in the *New York Times* headlined, 'HAVE FOUND HUSBANDS ABROAD: American Women Who Have Given Their Hearts and Money to Foreigners'. It began, 'It has been roughly estimated that English noblemen alone have captured by marriage with American women, in round numbers $50,000,000 of enviable American cash.'[3] In a way the Walpoles would doubtless consider most vulgar, the list attributed to most of the ladies specific amounts they apparently took to England, although Louise escaped that scrutiny. Nevertheless, it is most likely that Corbin's money played a significant part in her future mother-in-law's keenness to acquire her as a daughter-in-law.

Louise would discover that the widowed Laura Sophia Walpole was a formidable presence in the lives of her eldest son, his older sister, Amye Rachel (by marriage, the Viscountess Canterbury), and their younger brother, Clare Horatio, who, having quickly abandoned a naval career, joined the army, left that too, then married into an academic American family and lived in Virginia. Their Walpole ancestors had acquired vast estates in the heavily rural county of Norfolk in the east of England, where they built great mansions to reflect their status and wealth, most famously Houghton Hall, built in the 1720s for Prime Minister Sir Robert.[4]

Not to be outdone, Sir Robert's younger brother, Horatio – Dolly's four times great-grandfather – who was a diplomat and Whig politician, built another impressive Norfolk pile, Wolterton.[5] He also bought nearby Mannington, a thirteenth-century manor house with a moat and sunken gardens, while his nephew, Horace, the famed gothic author and Georgian man of letters, was building his white castellated fantasy, Strawberry Hill, in Twickenham, west London, which is still cherished by the public today.

As far as property was concerned, the family looked impressive on paper. However, when Louise and Walpole married in 1888, England was in the middle of an agricultural depression largely caused by competitive grain prices from the USA; ironically, the new accessibility of the prairies was helped by an expansion of railway building by people like D.C. Corbin. As a result, many of England's landowners found themselves asset rich but cash poor.

In addition, some of the Walpole houses had been desecrated by the owners themselves. Sir Robert's grandson, George, who inherited Houghton Hall, suffered from mental instability, fell badly into debt and sold its vast art collection to Catherine the Great of Russia.

Dolly's great-grandfather, Horatio, the 3rd Earl of Orford, liked to mess around with his surroundings. At Wolterton, he pulled down walls and changed windows, not always for the better, and replaced classic eighteenth-century furniture with modern Victorian, while disposing of artwork he disliked. He also hated the railways, preferring to make the long and uncomfortable carriage drive to London, and did everything he could to stop them coming near the house; his daughter, Lady Dorothy Nevill (Dolly's great-aunt), said it was 'a short-sighted policy and no good for the estate in the end'.[6] His son Frederick, Dolly's grandfather, who on leaving the Royal Navy became a Norfolk MP, bought a magnificent Elizabethan manor house, Rainthorpe Hall, where Dolly's father and his siblings were born; however, after Frederick's early death in 1876, his widow Laura sold the house and its contents, which included fine Jacobean and Chippendale furniture.

Nevertheless, when Louise and Walpole returned to England after their honeymoon, they had at least two residences available to them, in country and town: Waborne (or Weybourne) Hall in Norfolk, owned by his uncle who, after the end of his caper with the Countess of Lincoln was living in seclusion in London; and the house into which Dolly was born, 4 Queen's Gate Terrace in South Kensington.

As was customary, D.C. Corbin had set up a marriage settlement for his daughter; in doing so, he may also have had an eye to protecting her from Walpole's past. In November 1888, six months after their wedding, the hearing began at the High Court in London of a case brought against Walpole by a German governess, Valery Wiedemann. She was suing him for breach of promise of marriage and libel, for which she sought damages of £10,000. When Walpole met Louise, he may not have thought it necessary to share with her the details of his unfortunate liaison with Miss Wiedemann, for although she had first started court proceedings against him in 1883, after giving birth to a child she alleged was his, the case had lapsed. Besides, he and the governess had met in a faraway place when he was unattached. But if Louise did not already know all this when she agreed to marry him, she soon found out.

Wiedemann had reappeared in Walpole's orbit when she heard of his engagement. Between January and May 1888, she wrote to him, to D.C. Corbin and three times to Louise, telling her of the child, which she said was living. Wiedemann also defaced a picture of Louise from *LIFE* magazine and sent it to her, saying she was dishonourable for becoming engaged to a man who was promised to another. Her greatest wrath was reserved for Laura Walpole, who she blamed for causing her son to break his promise. In one of at least six postcards to her, Wiedemann wrote, 'You know that I must curse you from the bottom of my heart [...] for the endless suffering you have brought over me [...] I shall meet you once, and you will hear my curse.'[7] She also stalked Laura's house in Park Lane, London, and damaged the door, believing Walpole to be there.

As the hearing in November approached, the story was everywhere in the press, featuring as it did a member of such a well-known family. At the same time, it took a bold woman to sue for breach of promise, since her personal details would be bandied about in public. The undisputed facts were that Wiedemann (a good-looking woman a head taller than Walpole) had met him in 1881 in Constantinople, where she was working as a governess to the children of a hotel owner and Walpole was living on an allowance from his mother. They got to know each other and slept together. Beyond that, the facts were largely contentious.

The essence of her claim was that she had allowed herself to be seduced under a degree of violence and a promise of marriage and that Walpole had given her his signet ring (which was in her possession) and promised to take her to England and marry her within a few months. Walpole denied any such promise or violence, although did give her money so she could get to England.

Wiedemann went to Cannes, where Laura was staying, made herself known to her and, when it became clear that she would not entertain the idea of Wiedemann marrying her son, began to harass her. Walpole hired a private detective, Mr Cooke, to take Wiedemann to Paris. Her libel claim concerned what Walpole had said about her in his letter of instruction, which Cooke had foolishly shown her.

In that first hearing, Wiedemann was asked about the child she said was Walpole's but, against advice, she continuously avoided giving a straight answer as to whether it was alive or dead. The judge threatened her with contempt of court if she did not answer and gave her the option of retiring from the case, which she chose to do. As a result, the judge directed the jury to find in favour of the defendant.

Walpole barely had time to feel relief. Ten minutes later, Wiedemann's solicitor said she had not understood the implication of not answering the questions and they would be applying for a retrial. It was the start of a long-running nightmare for Walpole and Louise.

Such was the public sympathy for Wiedemann that a fund was set up to raise monies for her case, her champion being the London evening newspaper, the *Pall Mall Gazette*. Among letters it published offering money was one from 'a very poor governess – like Miss Wiedemann herself'. She said she knew:

> … what many of these 'upper circle' young men are like – rich, arrogant, overfed, and dissolute – and during the eighteen years I have been a governess, I have often met with young women who have been betrayed by men of this stamp, and cast adrift on the cold world with their reputation sullied – Heaven knows though it is not their fault![8]

It was even reported that Queen Victoria herself, on first hearing of Miss Wiedemann's misfortune, had sent her 25 guineas.[9]

The hearing to consider her application for a new trial – not even the trial itself – was set for May 1889, six months away. The humiliation of having her husband pilloried so publicly and the stress of having the case hanging over them must have put a great strain on Louise in what should have been a happy time, for she was expecting their first child in March, two months before the hearing.

No newspaper announced their daughter's birth, which was unusual, especially for aristocratic families, but doubtless it was to avoid bringing Dolly's existence to Wiedemann's attention. In April, at their church in South Kensington, the baby was christened Dorothy Rachel Melissa. Dorothy was an old family name; Rachel was for her father's aunt (the sister of Lady Dorothy Nevill), who had died too young; and Melissa was her mother's middle name.

Two weeks later at the court hearing, Wiedemann now admitted that her child had died shortly after birth. Her solicitor persuaded the court that she had not understood the consequences of refusing to answer the questions previously and a new trial was ordered. It did not take place until June 1890, over a year later, by which time Louise was two months pregnant with their second child.

To Walpole's dismay, Wiedemann was now representing herself, having not paid her solicitor. During the three-day hearing, Laura Walpole gave evidence for her son, saying that Wiedemann never mentioned any promise of marriage but had told her that she and Walpole were 'intimate'. Other witnesses included a count, who produced evidence that years earlier, Wiedemann had told him, falsely, that she had borne his child and tried to extract money from him.

While some attested to her good character, it lost credibility amid evidence to the contrary. This included a letter she sent Walpole in November 1884, saying he must keep his promise to marry her 'for the sake of our little boy', even though she knew the child was dead. The doctor who had attended her told the court he was born at eight months and did not survive.[10]

After contradicting herself many times, Wiedemann was found to be an unreliable witness. In the libel claim, the judge directed the jury that if Walpole believed the words he said to Cooke about her to be true, she would not be entitled to damages. The main question, though, was whether Walpole had promised her marriage. As the jury adjourned to consider its verdict, he had reason to feel cautiously optimistic. But when the foreman returned, he said they were so divided that a verdict could not be reached. The case would not be heard again for another year.

While Valery Wiedemann hung around like a miasma, it cannot have made Louise's acceptance by London society very easy, but friends stood by her. When Dolly was 6 months old, Louise stayed again at Inverlochy Castle and met the Danish artist Cathinca Amyot, who made a delightful pencil sketch of a little girl staring at her bespectacled nurse and asking her childish questions, as Dolly would do one day. She was too young yet to be aware of any family turmoil and spent much of her day in the care of her nurse.

As Walpole went between city and country, Louise's London base became an elegant and spacious property at 15 Grosvenor Square, Mayfair, probably leased by Corbin, whose wife and younger daughter were also living there. They employed twelve servants, four of whom were French – Mrs Corbin was accustomed to employing them and they were very fashionable – and one African. Louise certainly needed some help because on 9 January 1891, aged 24, she gave birth to a son, Horatio

Corbin Walpole. Now caution was thrown to the winds as his arrival was announced in several newspapers, with the words Walpole had been longing to write: the boy was 'the proximate heir to the Earldom of Orford'.

Whatever he dreaded might be the outcome of Wiedemann's case, his heart must have been lighter as the next hearing approached in June. His solicitor highlighted the inconsistencies in her evidence, while also acknowledging that the jury might not approve of some of Walpole's past conduct. Nevertheless, he said, it must not find that a promise of marriage was given.

But it did and awarded Wiedemann £300 in damages. Although that was far less than she had claimed, the verdict implicitly confirmed Walpole as a cad. Now it was his turn to appeal, on the grounds that there was no corroborative evidence to support her claim of breach of promise.[11] On 31 July 1891, all three court of appeal judges agreed and reversed the verdict in Walpole's favour, with costs – not that he was likely to see any, but it was a small price to pay for exoneration.

The fact that it had taken three hearings and nearly three years to reach this point was widely considered unacceptable, especially for Wiedemann, still seen by some as the victim.[12] But she had not finished yet. She threatened a claim against Laura Walpole for libel, then began giving classical recitals in London based on her experiences. Changing her approach, in 1892 she started selling pamphlets at the Liverpool Exchange berating the wrongs against her, before moving her spot to London and Royal Ascot. In a final shot, she sought permission to appeal the verdict against her as an impecunious litigant. She got no further.

Notwithstanding Wiedemann's continued attention seeking, the verdict in Walpole's favour had at least removed much of the strain and he and Louise could focus on enjoying their young children. However, such pleasure did not last for long. On 20 May 1893, at the age of 2 years and 4 months, Horatio died at Waborne Hall. Four days later, he was buried at Wickmere, the Walpole family church, with a large stone cross marking his little grave. In the church itself, in front of two luminous stained-glass windows of saints, a marble statue of a small boy with curly hair and angel's wings commemorates him.

Now Louise's daybook, a reminder of her carefree days of not so long ago, contained verses from 'Resignation' by the American poet

Henry W. Longfellow, about the death of a child, the sex adapted for their son. Written on black-bordered writing paper, three of the copied verses read:

He is not dead – the child of our affection –
But gone unto that school
Where he no longer needs our poor protection
And Christ himself doth rule.

Not as a child shall we again behold him
For when with raptures wild
In our embraces we again enfold him
He will not be a child.

But a fair youth in his Father's mansion
Clothed with celestial grace
And beautiful with all the soul's expansion
Shall we behold his face.[13]

It was written from 36 Bruton Street in Berkeley Square, the house Corbin had bought for her; undoubtedly anxious about all that had happened since his daughter's marriage, it was a way of further ensuring her security. She was still young, 26, when Horatio died and had time to produce another son, as her husband dearly hoped.

Dolly was 4 when she lost her brother – old enough to be aware not only of a sense of loss but of the grief of those around her. As if by osmosis, she began to absorb 'the ignominy of being "only a girl", that horrid phrase that haunted my extreme youth'.[14]

Female she may have been, but Dolly's education was not ignored, partly to endow her with the sort of skills and accomplishments expected of a girl of her background, with a view to making her desirable as a wife, and partly because of the Walpoles' intellectual heritage, whose legacy was in evidence among recent women. Dolly's great-aunt, Lady Dorothy Nevill, and her late sister, Rachel, despite the otherwise old-fashioned views of their father, had received from carefully chosen tutors an unusually wide education, which included being taught to read and write in Italian, Greek, French and Latin and was augmented with extensive travels in Europe.

Now, at the age of 67 and widowed for fifteen years, Dorothy Nevill was about to bring out a book on the history of the Walpoles and Mannington. It was bound to garner interest, for once the scandal of her early days was behind her, she had become a well-known figure in Victorian society, respected as much for her botanical knowledge (which won her the friendship of Charles Darwin) as her skills as a society hostess and confidante, especially to Benjamin Disraeli. Her views were still sought on a diversity of subjects, sometimes as ordinary as face soap, and she made no concessions to age, becoming endearingly eccentric in dress and demeanour. Circumstances would soon bring this clever, charismatic woman and her great-niece together. For now, though, the 'small and not particularly attractive' Dolly, who was 'rather lonely, rather shy and apologetic',[15] had dreams and invented games and began to read voraciously.

On 7 December 1894, Dolly's great uncle, the 4th Earl of Orford, died suddenly at his London house in Cavendish Square at the age of 82. It is unlikely she ever met him, for this cultivated, witty but eccentric nobleman, an old friend of Disraeli, had largely been a recluse for the past twenty-five years and separated from his wife for nearly forty, with their married daughters living abroad. As he left no legitimate male heir, his nephew Robert succeeded to the earldom and 5-year-old Dolly was elevated from 'miss' to 'lady'. Although the press extolled the deceased earl's intellect, it also raised his scandalous past, musing on how his career might have gone had it not been for his unhappy marriage and his eloping with the Countess of Lincoln. It was the same for Robert Walpole. He might now be the 5th Earl, but it would not be long before at least one newspaper slyly reminded its readers that he was perhaps best known for his court battle with Miss Wiedemann.

Dolly's life would change beyond her title, because her father inherited two Norfolk properties: Mannington, where the late earl was buried in a tomb he had erected within a ruined chapel in the grounds, and Wolterton, which he had abandoned years earlier in favour of Mannington. With these houses came chilling stories that must have thrilled young Dolly. The earl had had a premonition of his death, when he heard that a ghostly figure long associated with the Walpole family at Wolterton had been seen again: the White Lady, who was said to appear whenever calamity was about to befall them. One theory was that she was from the Scamler family, who formerly owned the estate, whose grave in Wolterton Park had been dug up by an earlier Walpole and caused her eternal unrest. Such was the fear of this apparition that the family devised the custom, whenever there was a death, of driving the hearse three times around the now-ruined church to atone for the act of sacrilege.

However, the late earl's sister, Lady Dorothy Nevill, who was raised at Wolterton, said she found no evidence of such an act and concluded that there was some other reason for her presence. Whatever the cause of this unhappy lady's state, a sighting always caused anxiety. Dorothy recorded that shortly before her brother died, he told her about the figure, saying ominously, 'It is you or I this time, for we are the only ones left.'[16]

As well as ghosts and graves, Robert Walpole – or to use the name by which he now signed himself, Orford – also inherited the contents of the houses, which included his uncle's magnificent library of rare books. The rest of the earl's estate was divided mostly between his illegitimate son and his sister, Dorothy: the only visitors he would receive in his seclusion.

As Wolterton was suffering from his neglect, its restoration was a long and expensive undertaking, and Dolly's parents made Mannington their main residence. A year after they moved in, Augustus Jessop, a clergyman and writer friend of Dorothy Nevill, told her:

> I don't hear much of Lord Orford and her ladyship. A friend of mine was at Mannington the other day and was greatly charmed with the daughter – her little ladyship (by way, Lady Dorothy the second) wrote the man a pretty little letter and he is so proud of it that I think he is going to frame it and hang it up in his private sanctum.[17]

At 6 years old, Dolly was already making an impression.

Like Wolterton, Mannington had much to feed the imagination of a bright and curious child. During Dr Jessop's own stay there in 1879, something happened that would fascinate firstly Norfolk and then the world. His vivid encounter in the library with a man thought to be Henry Walpole, a Jesuit who was executed in the sixteenth century, was a terrifying experience that he documented in great detail and remains a source of intrigue well over a century later.[18]

Smaller but older than Wolterton, the moated, medieval Mannington was beautiful and intriguing. Dolly loved playing in the grounds with its sacred groves, landscaped walks, sunken avenue of oaks, strange sepulchres and ivy-clad ruins. That which was not beautiful was curious. Above the entrance door was an eccentric reminder of the cynical nature of her late great-uncle in the form of an unpleasant testimonial to his wife and to his mistress, Lady Lincoln:

What is worse than a tigress? A demon.
What is worse than a demon? A woman.
What is worse than a woman? Nothing.

Dolly could ask her great-aunt more about the house, for her book on Mannington and the Walpoles was now published. The pair probably met for the first time when Lady Dorothy visited the new Earl and Countess of Orford shortly after her brother's death. As she told her friend Lady Airlie:

I had such a dear visit at my nephew's, such a delightful place with a moat of running water where Dorothy the 2nd fishes for our breakfast, and then of a night my nephew Robin [Orford's pet name] and I pored over all the books containing the sayings and doings of our ancestors – and then one day we went over to see lovely Blickling [another Norfolk estate] and its charming hostess – another day to a far-off estate that Robin has.[19]

Louise often invited her to dinner, showcasing her excellent French and knowledge of its cuisine in her handwritten menu cards. As Dolly grew older, she was sometimes invited to join them. Her great-aunt's attitude to life was one that would surely come to interest her. Dorothy Nevill disliked the fact that being a woman had stopped her from being accepted in a man's world and resented being told that to attend the all-male gatherings of her brilliant surgeon friend, the polymath Sir Henry Thompson, she would have to wear trousers. When Lord Lytton, on first meeting her, said, 'You are your brother in petticoats',[20] she took it as the highest compliment, holding the view that most women were stupid. Dolly no doubt sympathised with her frustration, but as she was born over half a century later, her stance would differ; while society was still patriarchal, women were (gradually) making inroads in areas dominated by men and Dolly would both celebrate and be an exemplar of the strengths of her sex.

Her own mother was an example of what a woman could do in the male-dominated world of sport. In 1895, Louise secured a place in the fishing annals as the only woman angler who had succeeded in catching two tarpon in one day when she, Orford and Dolly were holidaying in southern Florida the previous Christmas. One of the fish she caught weighed 128lb and although her husband caught a bigger specimen, he only managed the one.

For Dolly, that holiday also stood out because of her first sighting of an alligator. While her parents fished, she took advantage of her nurse's lapse of attention and wandered off along the riverbank. She stumbled across a party of men who had just landed what she saw as 'the dragon of all my fairy tales […] the ugly Apocryphal beast', whereupon she announced with all the grandiloquence of a 5-year-old, 'When I am big I shall kill those',[21] causing much laughter from the men whose prototypes would be her hunting companions one day. After all, she was only a girl.

2

A DIFFERENT DRUMMER

As they settled into Mannington, Wolterton was by no means forgotten because, as owner of the estate that covered seven parishes, Orford had wide responsibilities with which Louise assisted him. As lady of the manor, her participation in the community and support of charitable causes was expected and seems to have been generously given. In June 1897 alone, the earl and countess entertained 300 local schoolchildren at a tea party at Wolterton and he hosted dinner at a local hotel for all the male estate workers.

Dolly was also expected to get involved. At Christmas, she helped Louise with the annual distribution of flannel and calico to their cottage tenants, of which there were over 100. The amount of material each cottager received was calculated on how many children they had; if none, they received 3s instead. Each tenant also received a gift of coal from the earl.

For the time being, Wolterton's great house remained a faded glory which fascinated visitors, including Louise's sister, Mary, and her husband, Kirkland Cutter, a noted Spokane architect favoured by their father, who would replicate some of its features back in the USA.[1] Dolly's position as the only child remained, for her parents' wish for another had not been fulfilled. 'My little world became resigned to the fact that in due course I should have the succession of things,' she wrote. 'It was a goodly heritage, but of that I realised little at the time.' Without companions of her own age, she invented games which she played among the gardens and woodlands of Mannington and Wolterton and had dreams of 'pirates and savages, of desert islands and hairbreadth escapes'. They were, she concluded, 'premonitory dreams and games [...] unlike those of most little girls', and it was around this time that she began to hear 'the different drummer calling, one that sets a different beat to some of us [...] to do or be things that are accounted eccentric by more normal folk'.[2]

Her dreams often started in the same way. 'If I had a million pounds,' she would muse, she would conquer the world, climbing the Poles, exploring the solar system and passing 'the portals of eternity'. She made huts out of old packing cases, 'to the detriment of the shrubbery and the resentment of the gardeners', and hid in them any of her pocket money that had not been deducted for misbehaviour. From sticks and stones and bits of string she fashioned an armoury with which, crawling through the bushes and flower beds on all fours, she fought battles 'from the Crusades and the Armada to light skirmishing with howling cannibals'.

Although her parents sometimes took her travelling, they also spent time away on their own at favoured spots like the South of France, where friends kept a yacht and took them sailing in the Mediterranean. Like most upper-class children of the era, Dolly saw her parents for a limited time each day and was left in the care of servants. She made the most of it: 'While my governess placidly did her knitting and read love letters from her fiancé, a battered boat on the lake was my raking felucca,[3] with a skull and crossbones laboriously traced with ink on my pocket handkerchief.'

Without the distraction of siblings or other playmates, Dolly found magic in her surroundings. 'The park and gardens were haunted for me by goblins and djins; the donkey that drew my cart was an enchanted princess', all of which would influence her writing. Her story *The Call*,[4] which would be published in 1914, begins:

In the windswept garden, through the tormented February weather, the little white lady walked alone; such a very little lady, so young and slender and fragile – and always alone. She scarcely knew if she regretted it. She had known so little companionship; harsh, unfriendly parents, and now a still more harsh and unfriendly husband; for she was married, though she could hardly tell you herself how it had come about.

The lonely lady loves:

> ... the old neglected garden [...] with its maze of dusty tangled paths, its little dells of a damp, brilliant green and strange half-hidden moss-grown statues. And here her young soul grew. She had felt the mystery of every secret place; she had weaved a legend around every old stone image.

In the centre of a grassy spot, enshrined in laurels, she finds a statue of Pan whose artist had 'breathed into it some bizarre spirit of pagan days', and she is entranced by the 'virile slim lines of the body' and how the 'supple fingers' hold the pipe so lightly:

> It was here, on days when the sun shone that the little lady used to come and lie in the tangled grass and meditate on life, quaint, unusual meditations, nurtured on solitude and dreams.

The lady's dreams turn to erotic musings and, in her lonely state, Pan becomes her lover whom she visits at night, tiptoeing through the moonlit garden to rest her smooth cheek on the cold stone of his body. It cannot end well.

But for now, Dolly's thoughts were those of an innocent child excited by the books she read voraciously, most of which involved travel. *Swiss Family Robinson* was a favourite: a story of a family of immigrants who are shipwrecked en route to Australia; and *Arabian Nights*, the collection of ancient folk tales from the Middle East and Asia. She was absorbed by Robert Louis Stevenson's *Treasure Island* from which, to the despair of her governess, came her 'battle cry', 'Yo, ho, ho and a bottle of rum!' Added to these came fascinating tales of countries her parents had visited after their marriage – Japan, Ceylon (now known as Sri Lanka), the West Indies.

In March 1898 her parents sailed away on another extensive trip, which was to include British Columbia as well as their first visit to the

USA since they succeeded to their titles. As Louise was now a countess, the American press were eager to report her proposed visit to her home country: Corbin's main base was Spokane in Washington State, but he also had many business interests in New York.

Later that year, with her 24-year-old governess Kate Craven, Dolly joined her parents when they visited her Uncle Austin Corbin's 30,000-acre ranch in New Hampshire with its huge herd of buffalo, and they all spent the Christmas holidays in New York. She would remember her first sea journey as 'a very small and miserably seasick child, crossing the Atlantic to North America, in the teeth of an icy northerly gale',[5] although, fortunately, it did not put her off sea travel forever. She was becoming very fond of the USA, not just for its vast wildness but its human stories, of hold-ups and the pony post and the brave 'Forty-Niners', the thousands of hopefuls from all over the world who in 1849 made the long and often perilous journey to California to find their fortune in gold.

Back in England, the Orfords' lives followed a similar pattern to those of many upper-class families. Louise liked to spend the London season at her elegant house in Bruton Street, where she entertained friends such as fellow American William Waldorf Astor and took advantage of her father's gift of a box at the opera. She engaged in charity events, sometimes assisted by Dolly, while London's open spaces, especially Hyde Park, offered the opportunity for horse riding, at which mother and daughter were both accomplished; Louise was also known for her unusual but humane lack of the customary rein restraints on her horses. In a piece about American women in London in 1899, the *New York Times* noted, 'The Countess of Orford [...] has been driving a pair of American thoroughbreds in the park this week, the gift of her father Mr D.C. Corbin. The pair are pronounced the most perfect to be seen in London.'[6]

By contrast, the earl spent less time in London, preferring Norfolk, where he had many commitments and interests. In September that year, as the family settled into Mannington for the autumn, Orford hosted a shooting party at his hunting box at nearby Burnham Thorpe, birthplace of his ancestor Lord Nelson. Louise joined him and his guests at the lunch, where they posed for a photograph looking happy and relaxed. Generally, though, Louise found Norfolk too flat and preferred the dramatic scenery of Scotland, where the family spent three weeks every year in late summer.

In August 1900, Dolly's American grandmother died in Harrogate, West Yorkshire, her presence there undoubtedly for the purpose of taking the spa waters for her constant poor health. In a move that hardly seems conducive to good family relationships, Louise then joined her siblings in suing their father for half his estate, claiming as justification their mother's share under the property law of Washington State, where the Corbins' Spokane mansion was situated. However, Corbin avoided liability by arguing that he was actually resident in New York State, where he had lived and still had numerous business dealings, and where he said the bulk of his assets had accumulated. While his stance suggested parsimony, Louise could not be disappointed by her father's generosity to her, although Orford, feeling the economic effects of maintaining two estates, one of which he was starting to restore, complained that he seldom saw evidence of his father-in-law's wealth.

Before she reached her 12th birthday, Dolly lost her other grandmother too. Mrs Corbin's death was followed in January 1901 by that of Orford's mother, Laura Walpole, four days after mourning began for Queen Victoria. Laura's funeral was a curiously quiet affair, her eldest son and Louise being the main family mourners in the absence of her other children, Clare Horatio and Amye (Viscountess Canterbury).

Clare's lack of attendance might be explained on the grounds of distance as he lived in the USA, although he and his mother were not close. The reason for his sister's absence was an unhappy one. As Laura acknowledged in the 1897 codicil to her will, in the last couple of years, Amye had 'become of unsound mind';[7] she was 44 and that year had formally left the Church of England to be received by Rome (as her late uncle, the 4th Earl of Orford, had been). Whether she found that Catholicism helped her with her problems or whether the family considered her conversion itself to indicate a troubled mind cannot be determined. Either way, Laura optimistically made provision for her daughter in the event of her being able to manage her own affairs again, which had been put into the hands of trustees. In the meantime, she devolved responsibility to Orford, requesting him 'to do everything that may be in his power for the comfort and well being of his sister, my said daughter'.

Laura also made provision for her grandson, Henry Frederick Manners Sutton, Amye's son and heir to his father, the 4th Viscount Canterbury. She left him money on condition that he visit his mother 'frequently' and at least once in any half-year, together with a sum conditional upon his joining and continuing in the diplomatic or other 'honourable profession'. However, Henry, who was ten years older than his cousin Dolly, would not show much inclination towards anything respectable, and his ultimately short and dissolute life, during which he became 5th Viscount, would mostly be marked by his friendship with Oscar Wilde's lover, Lord Alfred Douglas, and with a wealthy German conman and a scandal involving a 14-year-old girl.

What Amye's husband was doing about her and why it was thought Orford should oversee her welfare must be surmised. Having inherited heavy debts, Viscount Canterbury had been declared bankrupt in 1888 and had reached an arrangement with his creditors; this, together with any anxiety about their son Henry, might explain Amye's state of mind. The viscount himself was in poor health, which would deteriorate quickly. Notwithstanding their mother's request, Orford had his sister committed to an asylum – a decision which the family today considers harsh, but perhaps he believed it was the best thing to do in the circumstances.[8]

Whether or not Dolly was aware of any dissonance among her father and his relations, her aunt Amye's seclusion and the loss of both her grandmothers meant that she saw her family diminish and still no sibling was forthcoming. After dealing with his mother's affairs, Orford and Louise went to Wiesbaden for his consultation with an ophthalmologist who had treated the late queen, leaving Dolly at Mannington with her governess and the family's Yorkshire terrier, Joey.

The following year, 1902, was a brighter one, in which the Orfords, particularly Louise, were increasingly enveloped in the new Edwardian society at a time when Americans were often considered vulgar arrivistes. As the first coronation in sixty-four years approached, they were commanded to attend a court at Buckingham Palace before King Edward VII and Queen Alexandra, where Louise's white satin dress embroidered in pearl and silver and decorated with old lace was one of the more beautiful gowns singled out for comment.

Dolly could see in her mother neither a social climber nor an aristocrat doing 'good works' to atone for her privilege, but a woman of diverse interests and commitments. In July, in a gesture that seemed to confirm the end of the Victorian era, the Conservative Prime Minister Robert Cecil, Marquess of Salisbury resigned due to ill health (to be succeeded by his nephew, Arthur Balfour), and gave a garden party at his country home, Hatfield House. Attended by the top echelons of the aristocracy, together with diplomats, politicians and a smattering of Indian princes, it was a major social event and Louise was there.

It was fitting she should be invited, for she was heavily involved in the Primrose League and had recently been elected Dame President of the Aylsham Habitation (as the area groups were called) in Norfolk. Named for the favourite flower of former Prime Minister Benjamin Disraeli, the League's aim was to spread Conservative principles by holding social activities and bringing politics to men and women who did not have the vote. Significantly, in those pre-suffrage days, it was the first time women had been given anything like a political voice.

Louise's involvement was no doubt encouraged by her husband's family. Orford's cousin, Sir Henry Drummond Wolff, had started the League and Lady Dorothy Nevill was one of its first female members and a dame. Her daughter Meresia had become its most active member, her interest fanned by her acquaintance with Disraeli through her mother's close friendship with him (which caused some people to ponder whether Meresia's brother Ralph was Disraeli's son, as they bore a marked similarity). Louise's commitment and Meresia's activism must have impressed Dolly with both illustrating what women in privileged positions could usefully do. Ultimately, the League was said to have furthered the suffrage cause, undoubtedly helped by female political influence in the higher echelons of society.

At the coronation (the date of which was hastily changed from 26 June to 9 August because of the king's appendicitis), Louise was one of several titled Americans who attended, the most prominent being the Duchess of Marlborough, the former Consuelo Vanderbilt, who was one of the queen's canopy bearers. Naturally, the US newspapers were keen to report on these fortunate ladies, although what Louise thought of observations that she 'wore fewer jewels than many present' must be imagined; however, those that did adorn her – a necklace of rubies, pearls and diamonds, and diamond brooches – were impressive. As she posed in her gown and countess's robe, bearing the weight of her coronet, the overall effect was one of elegant proportion, as her official photograph shows; a petite woman, she might easily have been swamped by too much fussiness. The earl also commissioned photographs of him and his wife and had them set in elaborate frames. Some newspapers said that Dolly, aged 13, was one of the few children to attend and was 'one of the most animated and interested spectators of the day'.[9]

As the daughter of nobility, she was finding she could be useful in her own right. She was enrolled as an Associate Member of the Children's Salon, a charitable organisation whose object was:

> To band together the Children of the Leisured Classes who have the wish and means to devote some of their time to competing with each other in such pursuits as Art, Literature, Music, Handicraft and Needlework, with the knowledge that all the work they do is sold to benefit some Charity especially those helping children less fortunate than themselves.[10]

Its patrons were a dazzling array of royals, headed by Queen Victoria's daughter, Princess Christian of Schleswig Holstein, and included young members of the royal families of Europe. A special section in *The Gentlewoman* was set aside for Children's Salon news and in March 1903, as she approached her 14th birthday, Dolly was the winner of the Associates' Art Prize (for the under 15s) for her design of a Children's Salon banner, which she executed in gold and pink on white in the prevailing Arts and Crafts style. There was an annual competition too, when members could meet up in London and compete on the day. To encourage entries (and recognising that the warm glow that came from helping a poor child might not be enough), there were attractive prizes such as silver-backed hairbrushes, silver-topped perfume bottles and manicure sets, awarded by eminent judges such as the actress Ellen Terry. In another competition, Dolly received an Honourable Mention for her watercolour of spring flowers; her writing would make its public appearance later.

Just a month later, she was giving out prizes herself on behalf of her mother, who at the last minute had gone to the USA, giving her health as the reason. Her intermittent problems were a shadow that hung over Dolly, although Louise's enthusiasm in all she did tended to disguise them. Orford did not mention this particular episode in his diary, recording only that she spent part of that year in the USA with her father.

At the Aylsham Industrial Society, an important Norfolk event showcasing local craftsmanship, much gratitude was formally expressed to Dolly for stepping in for the countess, and in responding, the earl reassured everyone that his wife's health was improving. Louise was now 37 and Orford still hoped a son was a possibility; his draft will of 31 March 1903 made provision for Louise to have an interest in his estate during her lifetime, after which it would pass to his sons and successive males. Daughters featured only when the men ran out.

By August, Louise was well enough to travel again to New York, this time taking Dolly with her. Sailing on the White Star Line's latest liner, *Oceanic*, they were part of an aristocratic contingent of sailing enthusiasts off to watch the America's Cup. Mother and daughter were guests of the Scottish-born grocery tycoon and sailing supremo, Sir Thomas Lipton, whose yacht *Shamrock III* they watched compete in New York City Harbor against the US team on *Reliance*, the eventual winner.

Louise's presence 'back home' was always remarked upon by the American press but praise of her qualities came almost at the expense of her husband's reputation. When Orford joined her and Dolly in October, at least one newspaper referred at length to the Wiedemann affair, even though it had ended twelve years earlier. Throughout the 'tempest of violent and extravagant abuse' to which the governess had subjected him, his wife 'stood loyally by his side and remained his best friend, perfectly indifferent and probably incredulous with regard to his conduct previous to her marriage'. Louise's reward came when his advisers showed Wiedemann to be 'a most dangerous adventuress and notorious blackmailer'.[11]

Now Dolly herself began to command attention as 'one of the most beautiful [how she must have rolled her eyes at that – vanity was not her failing] and accomplished girls in London's aristocratic juvenile set'. Wishing to emphasise her American credentials, its press talked of her friendship with Muriel White, daughter of diplomat Henry White, first secretary to the American Embassy in London, but while White and Louise were acquainted, Muriel was nine years older than Dolly, so 'friendship' was perhaps an overstatement. By contrast, the claim that Dolly 'delights in calling herself an American girl'[12] was probably correct, because she considered the USA her second home.

In their absence, Wolterton's ongoing refurbishment approached its final stage. On their return in early 1904, Louise continued to mastermind the aesthetics, but it took its toll and she fell ill once more. Summer fishing in Norway helped restore her, and under her expert tuition, Dolly improved her angling skills. By March 1905 they were able to move into Wolterton, the earl recording that, not only had the work taken a long time, it had cost him 'much money, not to mention time and thought, which however would have been a pleasure if the money position had been easier'.[13]

Much of the finance came from the sale of a Walpole property in Dean Street, London. It was worth it. Corbin visited them in February and could not fail to be impressed, as was Lady Dorothy Nevill on seeing her childhood home again. After 'fifty years of being allowed (the poor dear house) to go to the bats and the ravens', it was transformed. 'My nephew and Lady Orford (like most of her American countrywomen, enthusiastic about relics of the past) have done their best to restore the fine old Georgian mansion to its original state,'[14] she enthused. 'It really is one of the grandest houses I ever saw – seven receptions en suite – the bedrooms above, I had to take sixty stairs to my bedroom – they have made it most modernly comfortable. I missed nothing but the pen-wiper.'[15]

Mannington was rented out, although the first year of tenancy was marred by the death of a housemaid, 21-year-old Maud Roper, who was found drowned in the moat, whether by accident or by suicide the coroner was unable to determine.

Already familiar with Wolterton in its neglected state, Dolly found much to enjoy in its renaissance: its grand but light and airy rooms, her own suite and the family portraits, some of which Louise had managed to retrieve from places they had disappeared to. Having passed her 16th birthday, Dolly was beyond crawling in the undergrowth in search of adventure and instead learnt to drive (probably taught by the family's chauffeur in those days before compulsory testing). Nevertheless, Louise did not want her to grow up too quickly and became annoyed when it was said she was soon 'coming out'. The Countess of Orford apparently considered it 'a pity to introduce a girl [to Society] when she is under twenty' and, moreover, she should not marry 'until she is within measurable distance of five and twenty'.[16] Louise clearly relented on the first point because Dolly would come out at 18.

Now she was expected to assist more often with good works. 'It was subtly instilled into me that I had a part to play in the world,' she wrote.[17]

One of Louise's favourite causes was the animal charity, Our Dumb Friends League, for which she willingly embraced publicity. At one event, she was pictured smiling at the camera with other ladies outside a 'Gypsy caravan' selling craftwork, dressed supposedly as a 'smart Gypsy', while Dolly sat in a wicker chair wearing a folk costume and looking pointedly in the opposite direction.[18]

More attractive, of course, were the social invitations, and although Dolly was not yet invited on her own behalf, she was often included with her parents. Their circle included the Earl of Buckinghamshire, to whom Arthur Mills, Dolly's future husband, was related, and the wealthy manufacturing family, the Hope-Morleys, whose son Claude would marry Arthur's cousin. While Dolly is silent on how she and Arthur met, it is highly likely it was through such connections.

A significant shift in the family dynamic occurred in May 1906 when Orford's brother, Clare Horatio Walpole, died in Virginia aged just 48, leaving his widow, Ann, and their 21-year-old daughter, Amye, and while his death elicited only a fleeting mention in Louise's diary, it had potentially far-reaching consequences. Clare had been the heir presumptive to his older brother: if Orford died without a son, Clare would have become earl. While that meant the succession passing sideways rather than directly down Orford's line, at least the earldom would have passed to a close relative, as well as surviving a little longer. With Clare's death, the onus was more than ever on Louise to have a son, but the chances were becoming increasingly slim. However, at least some bonding came out of it, for when Clare's widow and daughter visited their relations in England later that year, Dolly and her cousin Amye would remain in touch for their lifetimes.

The year that would see Dolly coming out at last began in February 1907 with the start of the London season and a swish of society dances. The Windsor Ball was a lavish affair, patronised by royalty and held at the town's grand White Hart Hotel. Louise chaperoned her daughter, joining an interesting array of guests who danced until 3 a.m., an activity Dolly enjoyed and was good at. British nobility rubbed shoulders with American millionaires such as John Jacob Astor IV, who, five years later, would meet his tragic death on the *Titanic*, while Dolly's attention was bound to be drawn to the attractive coterie of army officers, some of them from the Coldstream and Grenadier Guards who protected the royal family at Windsor Castle.

Louise was keen to involve her daughter in other events as her 18th birthday approached, and they attended the opening of Parliament, with all its splendid pageantry, and a service in the Chapel Royal. At a time when it was customary to pay compliments to a young woman about to make her society debut, the American press was crushingly blunt. 'Lady Dorothy Walpole has a long face and the marked features of the Walpoles,' sneered *The Washington Post*, although it tried to redeem itself by adding that she has 'her mother's golden hair and dainty complexion. She combines the literary taste of her father with the sporting instincts of her mother.' In a sentence calculated to make the (assumed) husband-hunting debutante more attractive to her quarry, it ended, 'She will inherit vast wealth.'[19]

The timetable for the London season was dictated by the movements of the royal family. At Easter, Louise and Dolly returned to Norfolk, where they went to church and played golf at Blickling and Helmingham Hall in Suffolk, both important houses. Helmingham was moated like Mannington, but much larger, and shared a family connection. An earlier incumbent was Dolly's ancestor, Charlotte Walpole, the spirited granddaughter of Prime Minister Robert and the illegitimate daughter of his son, Edward. Aware of the stigma of illegitimacy, particularly for an unmarried woman, in 1760, without her father's knowledge or consent, Charlotte married Lionel Tollemache (later Earl Dysart), whose family owned Helmingham. She was an interesting player in what Dolly was coming to appreciate as her 'goodly heritage'.

The date for her presentation at court was set for 5 July, and although it was an occasion to look forward to, it was part of what Dolly would call 'the Juggernaut of family tradition'. In the preceding weeks, she and her mother went to operas and dinner parties, sporting events and galas, although in May their attention was diverted by the wedding in the USA of Louise's 75-year-old widowed father to his late wife's companion, Anna Larsen Petersen, a divorcée of 35. Of humble origins in Sweden, she had worked as the Corbins' housekeeper for twelve years, during which Mrs Corbin had educated her. It may have been with some wry amusement that the Earl of Orford, now 53, found himself with a new mother-in-law some eighteen years younger than he, although Louise was unlikely to be sanguine about her father's decision.

Not only was her stepmother her junior but her father had always kept rather aloof from Spokane society and his marriage to a former employee would cause gossip. Above all, she may have seen it as a potential threat to Corbin's estate, for while Louise enjoyed a benefit from the settlement he had made upon her marriage, he now had a young wife to consider, and his financial decisions might adversely affect his other children.

On 5 July, wearing the customary white dress and long gloves, with ostrich feathers in her hair, Dolly was presented by her mother to Edward VII and Queen Alexandra and her formal entrance into society began. Now she received invitations on her own account. 'For a while the world seemed mine to play with,' she wrote. 'I adored it all, the frocks, the parties, the dancing, the flirtations, the young men who sweated under the collar when they proposed and whom I had no intention of marrying.'[20] She enjoyed going to dances given for the debutantes who were presented with her, such as Princess Irene Duleep Singh, a daughter of the late Sir Duleep Singh, the last Maharajah of the Punjab. The princess's half-brother, Prince Frederick, who hosted her coming-out party at the Ritz, was another Norfolk resident.

The season culminated in a magnificent state ball at Buckingham Palace, whose invitation even Orford could not refuse. In the presence of the Prince and Princess of Wales and other royals, they danced the quadrille with Prince Frederick and Princess Irene, and if the main purpose of coming out was to find a husband, the princess would marry three years later.[21]

If Dolly was hoping for her own coming-out dance, it would have to wait a while, because after a strenuous season her mother was unwell again. In August, leaving Dolly and Orford at Wolterton, Louise checked into a seafront hotel at Margate and had a nurse to stay. It was not the usual haunt of aristocrats, being 'the resort of the cockney [...] overrun with excursionists from London',[22] but it was on medical advice.

In her absence, Dolly played hostess for her father and entertained his aunt, Lady Dorothy Nevill, and her niece, the Duchessa del Balzo, who lived in Naples. A charming photograph shows the three generations together, an occasion the most senior mentioned to a friend: 'We were three Lady Dorothys together at Wolterton, my niece Lady D. [who] married the Duca del Balzo, my great-niece, Lady D.W. and myself – most interesting.'[23] Perhaps Dolly mentioned proudly the publication of her elegant poem 'The Legend of the Buttercup' in *The Girl's Own Paper*,[24] a quality journal with a blend of educational articles and stories whose authors included the up-and-coming Angela Brazil and Noel Streatfeild.

Although Louise benefited from her seaside sojourn, on returning to Wolterton she was still weak and cancelled many engagements, as well as her usual winter visit to the USA. When the time came for the Orfords' annual shooting parties, Dolly managed with charm and self-possession the tricky task of helping to host them, which included keeping the ladies entertained while the men went shooting. She may have been assisted by Louise's younger friend, the Canadian-born Violet, Marchioness of Donegall, who stayed at Wolterton and whose elderly husband, the 5th Marquess, had recently died, leaving her with their young son. Despite her beauty and her charitable work, Violet was never properly accepted by society because she had answered the widowed and impecunious marquess' advertisement for an unmarried or widowed lady who was willing to marry him and pay for her peerage.

While the previous Christmas had seen Wolterton full of guests for several days, this year's was very quiet, no doubt in part because of preparation for Dolly's coming-out party on 27 December. Over 300 guests came, her parents' acquaintances bringing with them their young adult offspring and their friends and a flurry of titles. Perhaps there were romantic hopes (at least on the part of the young man and his parents) for Dolly and Geoffrey Hope-Morley, from the manufacturing family the Orfords knew well. Maybe Captain Charles Henry Murray St Clair, a courageous army officer and heir to his father's baronetcy, hoped to find the woman of his dreams at the ball – he had not yet had time to do so, as he had been busy fighting with distinction in the Boer War. As for the debutante in whose honour the ball was held, she loved the dancing and the flirting, making up in wit and humour what she lacked in conventional beauty, while also developing a keen eye for social observation.

If any of those young people who danced so gaily on that happy evening, or their elders who watched so fondly, could have glimpsed the future, what might they have seen? That Charles St Clair never succeeded to his father's title because he would be killed in December 1914 at the Battle of Givenchy. That Geoffrey Hope-Morley would have three wives, none of whom was Dolly. And that she would find love with a man outside familial expectations, while also discovering that peace lay not in the English shires but in a very different part of the world. For now, though, in the heady excitement of the moment, her dreams of the worlds she wanted to explore were put aside, the call of the drummer silenced.

THE DEBUTANTE

Dolly was becoming particularly interested in the way women were regarded by society, a subject she would keenly observe in other cultures she encountered. When her relation tactlessly remarked on her need to be interesting in the absence of beauty, she implied that the two qualities were mutually exclusive and if only one were possible, then good looks would always win – the 'fascination factor' kicked in only if one had lived a remarkable life.

When her great-aunt Lady Dorothy Nevill turned 81 in 1908, *The Tatler* published a full-page photograph and called her 'one of the most interesting personalities living [...] the last of the *grandes dames*'.[1] By contrast, however clever or fascinating a young woman might be, the social pages usually focused on her appearance. Dolly accepted there were benefits to beauty but considered it a rare quality, believing instead in the power of body awareness. 'Plain women would no longer be plain if they had poise,' she said in an interview. To achieve that, 'No clothing which confines the body too closely must be worn'.[2]

In 1908, a style of clothing that was wriggling its way free of earlier restraints began to appear, and 19-year-old Dolly, whose childhood shyness was now replaced (or perhaps disguised) by a fondness for unusual, even theatrical dressing, seized the moment. Embracing the fluidity and simplicity of the new fashion, her chosen gown for a state ball (in honour of the French president) was striking enough to merit wide comment. Its very simplicity gave it 'an enviable cachet', crucially without losing its smartness,[3] and was relieved by delicate embroidery of fine gold thread and gold beads in a classical Greek key pattern. Dolly was influenced by the ancient Greeks and their approach to exercise and movement, and she may also have taken inspiration from the new phenomenon, the performer Isadora Duncan, whose controversial free style saw the birth of modern dance. Her sense of style was attractive and her humour made her good company.

After spending the start of the year with her parents in St Moritz, whose warm climate was good for Louise's health, Dolly was thrust into her second London season. Entertainment came in many forms, the most interesting naturally being those where she met other young people, for which dinner and house party invitations from family friends, such as the Coats, could be useful. They were wealthy industrialists, whose long-standing family business making cotton thread had become one of the largest companies in the world.

Sporting events were popular with Dolly and her mother, especially if horses were involved. As well as attending the near-obligatory Royal Ascot, she and both parents joined the Hope-Morleys in their carriage for the Four-in-Hand Club event in Hyde Park, one of the greatest attractions of the season. Gentleman drivers who were members of the old-established club drove their own four-horse coaches in a procession through the park and surrounding streets, their elegant vehicles and elite passengers drawing large crowds and necessitating a police presence.

That year, a unique attraction came in the form of the Olympic Games, unexpectedly hosted by London: they were supposed to take place in Rome until various factors in Italy made that unworkable. They were the first to be organised by the sporting bodies concerned and the first to have an opening ceremony, but the parade of athletes, like the games themselves, was marred by politics and controversy, including the refusal of one of Louise's countrymen to dip the US flag in salute to Edward VII, a practice that continued.

While the London season was primarily about socialising, there was often a political or charitable cause behind a gathering. Orford and Louise were members of the London Municipal Society, which had been recently formed to promote the interests of the Conservative Party in local politics. An invitation to the society's garden party at

Strawberry Hill, home of Orford's ancestor, Horace Walpole, was too good to miss, combining as it did both a chance to chat to former Prime Minister Arthur Balfour and to see what the current owners, the supremely wealthy Lord and Lady Michelham, were doing with the gothic mansion.

Most important to Louise was the grand country fair held in Regent's Park in aid of Our Dumb Friends League, for which her committee succeeded in obtaining the patronage of the Prince and Princess of Wales (the future George V and Queen Mary). Meanwhile, Dolly was involved in her own good cause, helping to organise a charity concert in London in aid of the terminally ill.

In August, Dolly and Orford left for Scotland while Louise went to New York to visit her father. As the year drew to a close and the family greeted the next, with mother and daughter attending a lively ball before leaving again for St Moritz, it looked as if 1909 would follow the same pattern as previous years. But on 4 May, life changed forever.

At Wolterton, as Louise was preparing to go for a ride in their motor car, she collapsed in her room. Her maid hastily summoned Dolly – Orford was not at home – but it was too late. At 42, her killer was heart disease. The earl recorded her death in his estate diary, an entry which was short and factual rather than emotional, slotting in as it did between accounts of tree planting and crop rotation: 'This year on May 4th Lady Orford suddenly fell dead in her bedroom (the state bedroom). I was on my way to Scotland at the time.'[4] An album was dedicated to dozens of newspaper cuttings about the countess's death, one of which mentioned her having restored Wolterton and replaced many of its lost treasures. Across the bottom her husband added a comment: 'NB. Lady Orford never spent any money on the restoration of Wolterton, it was entirely done at my expense.'[5]

The Countess of Orford was buried at Wickmere in a simple service, her peeress's coronet carried on a velvet cushion alongside her coffin. Mourners included estate workers, heads of charities and local schoolchildren, whose posy of primroses was left on the coffin lid as she was lowered into a grave next to that of her little son, Horatio.

Louise had impressed people not only with her combination of intellect and a love of the outdoors, but with her generosity of spirit. Now some of her community duties devolved to 20-year-old Dolly, who as the unmarried daughter also had the task of being companion to her 55-year-old father – 'Lord Orford's only comfort', as one newspaper put it – without any siblings to share it with.

Her mother had been anxious that she should not marry too early, but the earl, being more of a traditionalist and aware that Dolly was so often the bridesmaid rather than the bride, thought differently. Even their staff were getting wed. Charmingly, in January 1910 at Wickmere Church, Orford's butler, Edward, married Dolly's French maid, Charlotte, for which Dolly provided the bride's veil and wreath of orange blossom.

That month, Orford left Dolly in London with the Hope-Morleys at their house in Grosvenor Square, closed up Wolterton and went to New Zealand for five months with a local friend, Sir Reginald Beauchamp, who had also lost his wife, this time to another man. In her father's absence, she wrote more poetry and took part in a masque for charity in the grounds of Blickling Hall with a classical Greek theme, of which was said admiringly, 'Lady Dorothy Walpole made a singularly graceful and silver-spoken Aphrodite'.[6]

On his return, Orford resumed his social diary: as an aristocratic widower with a clever daughter of marriageable age, they were not short of invitations. A house party at Holkham Hall given by the Earl and Countess of Leicester, for example, included their unmarried son and other bachelors, as well as the widowed Cora, Countess of Strafford, whose beauty had recently been captured in a portrait by John Singer Sargent.

Dolly's autobiography offers little detail about this stage of her life, including her mother's death. It was said of her that she mixed only with 'the cultured and intellectual set. Of the smart set she fights shy; wild horses won't draw her near them, though they bombard her with invitations.'[7] Given the light in which she would portray the 'smart set' in her novels, this is probably quite accurate.

As for her romantic life, according to 'friends', 'a famous man of letters, part proprietor and editor of a high-class weekly, has been paying her considerable attention [...] She herself explains their great friendship by the fact that they have so much in common. He is fifteen years older than she.' Tantalisingly, there were a few impressive journalistic figures at that time, although none is an obvious candidate. Her poem 'Why?', composed (fittingly, in Venus and Adonis stanzas) in around 1909, is about a romantic experience which was surely her own:

> A year ago you gathered me a rose,
>   With some light heedless compliment, and I –
> I loved you then. I took it, with God knows
>   What wild, tumultuous heartbeats. 'Till I die' –
> I thought. 'I'll keep it.' With such pulsing gladness
>   This trifle shook my soul almost to madness!
>
> Last night, before my feet you madly cast
>   Your heart, yourself, your All; and I (I do
> Not love you now, in spite of all the past) –
>   I threw them from me, lightly mocking you.
> And now no plea or prayer that you can fashion
>   Can move my heart, or even my compassion![8]

While the earl had pursued his own agenda earlier in the year – he had fallen in love with New Zealand while in the Royal Navy and now sought a property there – he had also planned a world trip for them on his return, perhaps to mark Dolly's 21st birthday, which he had missed. They left for Nova Scotia in August and were given the use of Corbin's private car to travel through Canada, then visited him in Spokane. From there, they went south to California, where Dolly achieved fame on Santa Catalina Island by shooting a wild goat at 500 yards without leaving their vehicle. The feat was reported with excitement by the *Los Angeles Times*, which also enthused about her angling skills, surely honed under her mother's tuition: she had been seen tackling a yellowtail, a large and powerful fish, for nearly two hours. Mexico was also on their agenda, followed later by New Zealand and Australia. Their return voyage included Shanghai, Penang, Kolkata and Aden, arriving home in April 1911 in time for the London season.

Louise's house in Bruton Street had been sold, so for the 1911 season they stayed instead in a rented flat in Buckingham Palace Mansions, a suitably central location for a major celebration in June, the coronation of George V and Queen Mary. As such, there were more dances and parties than usual, aligned as they were with coronation festivities. A more serious marker of the event was the Imperial Conference, one of an occasional series attended by representatives of the autonomous communities of the Empire (later called the Commonwealth). That year, the focus was on proposals made by New Zealand's prime minister and, as an indication of how important that country was becoming to him, Orford invited his representative and other officials to Wolterton, where Dolly acted as hostess.

She was now labelled 'the most travelled young woman of her age in the kingdom'. More prosaically, it was also noted that she would be 'seen as a bridesmaid at several of the early autumn weddings'.[9] At 22, perhaps it was thought she might become an *old* maid, although she was aware of expectations: 'I knew that some day I had to make "a good match", a prospect that I classed alongside a visit to the dentist.'[10]

Nevertheless, she complied with the custom of being photographed as a debutante for the society pages, for which some of the best professionals were female. For now, it was Lallie Charles, whose studio portrait of Dolly wearing furs and an extravagant hat appeared in *The Tatler* in January 1912, the wording mostly putting her in the context of her parents: 'Lord Orford's only daughter […] is endowed with much of that personal charm for which her mother was so famous.' The timing coincided with her attending the Bachelors' Ball at Lewes in Sussex (perhaps 'encouraged' by her father), an annual event whose theme was as obvious as its title suggested, where she joined 300 other young men and women in dancing to a Viennese orchestra.

Two months in Italy with her father followed, where they visited Naples, Florence, Rome and Venice. Always an enthusiastic buyer of *objets d'art*, the earl keenly recorded various items and noted that 'Dolly and I had our busts done in marble for £55 each' by a Florentine sculptor.[11]

Later that year, at the Coats' house party in Scotland, they were reminded of the good match their hosts' daughter Maud had made three years earlier in marrying Arthur Wellesley, the future 5th Duke of Wellington. Dolly's poem 'Doubts', published in March 1913, suggests there was someone who mattered to her, even if she was conflicted about it:

Pure and saintly eyes beneath a thoughtful brow
Passionate red mouth where longing mutely lies
– Oh my Heart's desire, what doubts beset me now
As I gaze upon your lips and in your eyes![12]

An element of cynicism also pervades her writing, seen in another poem published the same month. 'A Type' is surely a metaphor for the relationship between the sexes. A butterfly is 'Caught in the web that men will weave/For charming playthings just as these'. Its beauty and fragility appeal but then man crushes and destroys it. It cannot be saved, 'since Fate will not defend/Such pretty useless things'. The poem reflects Dolly's keen observation of the sexes, a subject on which she would write profusely, as she would on the foibles of human nature.

Her disdain for superficiality, such as the importance placed on a woman's appearance, is evident in her short story, *A Fable*, published in September 1913, which looks at the poisoned chalice that is beauty. Beauty visits Earth with her child, Love, and afterwards tells Zeus, 'With the humans, Beauty is not a goddess. She is to all but a piece of merchandise and to most an object of suspicion.'[13]

According to her autobiography, Dolly's marriage to Captain Arthur Mills in 1916 came after three years of 'family warfare', suggesting that they had met or fallen in love around 1913, although it is not easy to see how Arthur fitted into her life at that stage. Whenever it was that they first met, it was almost certainly through shared connections. Perhaps Arthur was the subject of 'Doubts', although at that time a rumour was circulating about Dolly and a young Life Guardsman (which Arthur was not) and her continued attachment to a rosette of pink silk ribbon.[14] She had carried such an item at a wedding in June that year at which she was a bridesmaid and the best man was 29-year-old Captain Gosselin of the Grenadier Guards, the son of a distinguished diplomat. If there were any grounds for deducing feelings on Dolly's part, it is as well that nothing came of it, for he would be killed in action two years later.

The uncertainty about her and Arthur is exacerbated by a newspaper report in October 1913 whose byline read, 'Lady Dorothy Walpole Announces that She's Resolved to Be an Old Maid', which quoted her as saying:

I am in love with my writing and I want to give up my life to it. I would not marry the proverbial archangel. If I were to marry I would have to devote myself to a man and my heart would be all the time in my work.

She also reportedly dressed in 'an elderly manner' which she said was 'to keep the young men off' so that they would see she was 'not a marrying girl'.[15] Tired of speculation about her love life, perhaps she was having a joke or even publicly denying any attachment to appease her father, because whenever it was that she and Arthur fell in love, Orford did not approve of him as a prospective son-in-law.

Arthur had everything, wrote Dolly, except money, and it seems that was the problem, despite his being the nephew of the Earl of Buckinghamshire and the son of a respected church minister, whose appointments included Assistant Chaplain of the Chapel Royal. Arthur's stepmother, Elizabeth, whom his widowed father married when his son was about 7, was from Scottish aristocracy. But Orford's aim, like that of most upper-class fathers, was to marry his daughter to a man who could provide for her in no less a manner than she had been accustomed to, and if he doubted that Arthur, a career soldier, could do so then his anxiety was understandable.

Dolly also said the advent of the Great War complicated matters. Arthur was sent home in October 1914 after being wounded at La Bassée, so her father may have feared he would not be fit for employment afterwards, or if he did return to action, that Dolly may be widowed. Any anxiety the earl felt about her vulnerability if she married a serving officer would not have diminished with the news of the death in action of a relation at the Battle of Loos in September 1915. 'Thomas [Henry Bourke Vade-] Walpole, my cousin and heir presumptive to my two baronetcies, was shot dead in the trenches,' he recorded.[16] His death also raised, yet again, the issue of succession. In the absence of a son for the earl, Lieutenant Thomas Vade-Walpole had been in line to inherit two other titles the earl held, the Baronetcies of Walpole and Wolterton. Now Thomas' position in the line of succession shifted to his younger brother, Horatio.

Whatever his reasons for taking against Arthur, Orford's apparent disregard for Dolly's feelings appears harsh, while his threat of disinheriting her hardly seems the act of a loving father. Perhaps there were other reasons for his antagonism. By its very nature, Dolly's autobiography is subjective. While money may well have been at the root of her father's attitude, perhaps there was something else about Arthur to which he objected that Dolly was reluctant to admit. Whatever it was, it would have to be sufficiently serious to make Orford willing to alienate his only child.

By the time she and Arthur married in June 1916, she must have hoped her father would relent, for it seems there was much about her husband to admire. He had not wasted time while recovering from his injury and had two non-fiction books in print under the pseudonym 'Platoon Commander', *With my Regiment: From the Aisne to La Bassée*, already in its second edition, followed by *Hospital Days*. Combining grim reality with gritty humour, both were well received.

While the earl's diary was mostly concerned with estate matters rather than the personal, his entry about the wedding nevertheless feels brusque in its brevity: 'My daughter Lady D got married in June.' However, she had more pleasant things than family feuds to think about as she joined Arthur at the altar. Worn out by the organisation, which she later described as being harder than running an African safari, as she walked up the aisle 'the flowers and the people and the strains of sweet music seemed to whirl around me in a mist'.[17]

The absence of her parents – her mother through death, her father through choice – made the day bitter-sweet. Another familiar face that was missing was that of her great-aunt, for Lady Dorothy Nevill had died in 1913 (an event which elicited personal responses from the king and queen).

In the church decorated with red roses and lilies, they were married by the sub-dean of the Chapel Royal, a colleague of her father-in-law, who was assisted by a figure familiar to Dolly, the Rector of Wickmere. Afterwards, they drove away for their honeymoon on the south coast, her chic blue and white outfit, like her gold wedding gown, giving no hint of their lack of funds.

Beyond the wedding, there was the exciting anticipation of Dolly's first book, her novel *Card Houses*, which was to be published later that year. In the 'sprightly society skit', pleasure-loving Cleo Stayres is 'the penniless daughter of an indifferent baronet', who gads about until an uncle takes her on a world tour and she finds change and romance and ultimately marriage. Reviews varied. *The Sketch* posited that it was written to a 'certain very famous recipe' originally found in a 'brilliant' essay called *The Decay of Lying*, although it nevertheless invited its readers to enjoy the book.[18] *The Gentlewoman* said the tale was 'too fragile to stand the weight of detailed criticism',[19] but while Dolly would be the first to admit it was not great literature, there was, as there always has been, a ready audience for light reading, especially when it describes a world outside the reader's experience.

The novel-writing process was a learning curve for Dolly. Given that it drew on 'many well-known society people [who] find themselves portrayed under lightly veiled personalities',[20] there were consequences. 'My first novel nearly got me into the Divorce Courts,' she wrote, 'because an unintentionally too-true pen picture of a certain Guardsman made a jealous wife imagine that I had really spent a weekend with him in Paris.'[21]

As Dolly tried to settle into married life, the disadvantages of her privileged upbringing soon became apparent. Not only was she unaccustomed to doing anything for herself, having always relied on a maid, but she had 'no knowledge of housekeeping in any shape or form'. As a member of the pre-war British Army, Arthur's service in France so early in the conflict was rewarded in 1917 with the medal known as the Mons Star[22] and he returned to war to serve in the Palestine campaign. Conscious of her own 'general uselessness', especially when other young women were doing 'heroic' things, she took matters in hand. As well as doing voluntary work in London's poorest area, the East End, she opened a 'gift house' in North Audley Street in 1918 in aid of the Bulldog Club for Sailors and Soldiers, for which she invited quality donations that could be sold at market prices. To earn her own money, when she had the time and energy, she turned to writing.

Not only was she living without Arthur, but her father had surprised, even shocked her, when in September 1917, at the age of 63, he married 25-year-old Emily Gladys Oakes. Dolly did not attend the wedding, which took place in Hampshire, where Gladys' father, who gave her away, had been vicar. A Walpole relation who was the Bishop of

Edinburgh conducted the service,[23] with the loyal Rector of Wickmere assisting once more. Far from being the quiet wedding anticipated, it aroused so much local interest that a police presence was required, and little wonder. Not only was there a big difference in age but in social status too: 'Peer weds vicar's daughter,' announced the press.

If the widowed earl wanted mostly companionship, he could have found it with a woman nearer his age, but he wanted a male heir and it fell to this pretty young woman, now the Countess of Orford, to provide it. Immediately after the ceremony, they left for Wolterton, her new home. This time, his diary entry was more effusive, even detailing their meeting in Argyllshire where he had bought a sporting estate the previous year.[24]

A year after the wedding, 'My dear wife gave me a daughter born at Netty St Matthew [her family home],' the earl positively gushed. The arrival of Lady Gladys Grace Sophia Walpole, 'born at 4.30pm on Wednesday Sept 4th 1918, weight 6lb 1oz',[25] was clearly a joy, although a comment in the press brutally encapsulated the patriarchal view which Orford, if pushed, might have admitted. 'New additions to the peer-age have come thick and fast of late,' said *The Sketch*. 'Lady Orford's small daughter must, one can't help thinking, have brought a tinge of disappointment to her father, who still remains without an heir to the Earldom.'[26] Although the child was Dolly's half-sister, it seems unlikely she saw her, and she is not named among the guests at the christening at Wolterton a month later.

Exactly how Dolly viewed her father's wife must be surmised. While technically she was her stepmother, applying that term to a younger woman must surely have seemed unnatural.

Less than a year after their marriage, Dolly lost her grandfather. On 29 June 1918, D.C. Corbin died in Spokane, aged 86. He would be remembered as 'a major shaper of the growth and prosperity of Spokane, the economic and geographic centre of the Inland Northwest'.[27] He had had Dolly's future in mind from the start, his will stipulating:

As to my granddaughter, Lady Dorothy Walpole, daughter of my deceased daughter Louise, Countess of Orford, I make no allowance to her [in this will] for the reason that at the death of her father she will inherit and come into by the terms of a settlement made by me at the time of her mother's marriage a sum now held in trust by the General Trust Company of New York City, yielding an income of two thousand pounds sterling per annum.[28]

Although Dolly was not born when he created the marriage settlement, clearly Corbin, with an eye on his future son-in-law, felt it prudent to provide for any children. However, as Orford had become a father again at the age of 64, there was little sign of his imminent demise. She would have to wait.

Meanwhile, there were signs that the war might be coming to an end. When the bulletins were more reassuring, Dolly:

> … mingled with the unhappy, hectic crowd that in dancing and noise tried to kill an ever-present gnawing anxiety. Much has been said against these war parties but I learnt then that often they were the ultimate buffer against despair, a safety valve from recklessness and suicide.[29]

Peace was declared in November 1918. Arthur returned home, and adjustments had to be made. He resigned his commission and concentrated on his literary career, but it was early days, and he was making little money. 'Romance,' Dolly wrote, 'had come down to earth.'

She realised then, as perhaps she had always known, that she could not sit back and simply be someone's wife. She had to do something of her own. 'I plumped for literature, and I *worked*. I think that before I found my literary feet, I ran almost the entire journalistic gamut.' She took anything that was offered to her, at any price, and that approach, together with the cachet that came with her heritage and her title, meant she was not short of offers; nothing, however, could guarantee success.

A fortnightly paper hired her to write 'causeries', the concept of which she never quite understood, but she did it for six months and then got the sack, which made her weep, although not as much as thinking of the 'miserable pittance' they had paid her. She reported prize fights, wrote book reviews and articles on any topic that was suggested to her.

She wrote society paragraphs too, but they could get her into trouble: the first job she took only lasted two weeks, at the end of which her 'justly enraged boss' told her she had let his paper in for three libel actions. But her articles themselves were lively and aimed at the Modern Woman, who was emerging from the altered landscape of the Great War. As well as undertaking demanding voluntary work, many women had become wage earners, often doing men's work, and were developing an awareness of their economic worth. After decades of campaigning, the Representation of the People Act 1918 gave (some) women the right to vote. From an increased sense of freedom an array of female interests were emerging, to be covered by lively writers like Dolly.

Although she could already drive, in 1919 there were fewer than 1 million drivers in the UK. Those women who owned a car or aspired to do so might have been interested in Dolly's recommendation in her contribution to a column called 'Woman & her Car' (written by a man). Her favourite was 'the 15 hp. Silent Knight Daimler. It is of convenient size for London use,' she advised, 'and can do up to 50 miles on a country road. It is moderate in price, easy to drive and extremely comfortable.'[30] Women who enjoyed country driving could have their experience 'doubly sweetened', said the column, and do their bit for the post-war effort by taking with them 'a wounded officer or "Tommy" to whom they would give fresh air and aid to their convalescence which would quicken recovery immeasurably'. Perhaps the organiser of this charitable scheme (Madame de Latour of Knightsbridge) hoped that rather than the warm feeling of doing good, the new shortage of men would be the persuading factor in getting a woman to drive with a stranger into the countryside.

After the war, a sense of frenzy seemed to take hold of the younger generation from whom so many young men had been taken. Jazz was on the rise, finding its way into the emerging nightclubs, but it was viewed with suspicion by many. Dolly loved it.

In May 1919, she was the subject of a 'jazz portrait', drawn by Alfred John Bennett, an ink drawing which captures her in profile wearing a jazzy-looking dress and headband; it would later be exhibited at a London gallery. In a piece headlined 'Why I Jazz – Because I Like It!' she explains that the term does not really exist on its own, except to denote a type of dancing and is done to jazz *music*. She had not heard of the term until someone said to her, 'How divinely you jazz!' To jazz 'requires the maximum of effect with the minimum of effort' and has huge physical and mental benefits, she enthused. 'Dancing keeps one fit and sound [...] The inspiriting din and infectious rhythm of the music banishes fatigue and goes to one's head, as does champagne – without the subsequent headache.'[31] As for those who considered it immoral, it was as modest or dignified 'as a frock or a funny story might be – dependent on the personality of its perpetrator. Some people could make anything vulgar!' Jazzing meant that 'For the time being life is reduced to its simplest proportions', a notion that would increasingly appeal to her.

With cinema-going becoming increasingly popular, Dolly wrote vividly about her experience of being a film extra in a London studio, where she was 'a grand duchess or something unimportant in the background'. She made a mess of applying her make-up – 'every vestige of natural flesh tints has to be thickly painted out with yellow grease paint and powder' – and was told off for laughing and spoiling a scene.[32] In watching the leading actors, a pretty woman and an 'extremely beautiful and romantic young man', she realised she did not have the face for 'kinema' (as it was called) but thought it just as well.

This was still the time of silent movies, and the leads made acting look easy, 'until one's own turn came and one realised the difficulty of expressing, bodily and facially, every kind of violent emotion without the help of ready-made words'. She was touched by the good nature of the cast towards the extras, who were often those with:

… forlorn hopes [...] those who have failed to make good on the real stage [...] men whose voices or health have failed them; women who are too old or whom circumstances have thrown on the world – all find their way to the kinema. There is room for them all.

Dolly had also written her second novel *The Laughter of Fools*, to be published by Duckworth the following year.

While she was busy with her writing and occupied with more charity work – like her mother, she was a good organiser and sat on various committees – in June 1919, her half-sister, Sophia, died, at the age of 9 months. 'Strong and healthy, she was suddenly taken from us, the cause was meningitis,' wrote her bereft father.[33] The countess was already expecting another child and on 11 December gave birth to Lady Anne Sophia Walpole. Again, the press could not resist mentioning the lack of an heir to the earldom. Even the heir presumptive to the baronetcies had changed once more – in 1918, Horatio Vade-Walpole was killed in action like his brother; his position passed to his 6-year-old son Robert, known as Bobby. How Orford must have thought with heavy heart of his own son he had lost.

Lady Anne Walpole was christened on 10 January 1920. In the unlikely event that Dolly was invited, she would have been unable to attend as she had already left on her travels to start a new phase of her life.

THE CALL OF THE WILD

In December 1919, as *The Laughter of Fools* was nearing publication, Dolly left for North Africa on a journey that would mark the start of a love affair. She wrote:

> After three years of marriage, an accident of ill-health sent me south on to a warm climate that happened to be Algiers, and to Biskra and beyond, and I found the Sahara and the great wild places. Or rather, they found me, and took me in a relentless grasp that to this day has never weakened.[1]

In the wilds she would find 'realisation of all the dreams and fantastic games of my childhood' and, above all, 'peace and happiness and rest, and a respite from all the underlying problems of life'.

Her privileged background could not protect her from life's vicissitudes. The war had taken its toll, as it had on countless others. She felt alienated from her father and his new family. As much as she loved Arthur, marriage demanded compromises. To be able to lose herself in a different existence would be her salvation.

Her choice of the word 'respite' – a short period of rest or relief – is telling. However much she needed the wilderness, she could never entirely give up her world. She was interested in people. She loved dancing and dressing up. To smoke a cigarette and quaff a cocktail was a joy. And she could not turn her back on her heritage. All this would draw her back to England every year.

Dolly was unlikely to deny that her travels were motivated not only by curiosity but by self-preservation. Fortunately for posterity, however, not only did she have the passion, even obsession, of the explorer but she found ambition. 'I had found my *métier* as a writer, for I am one of those who writes with the heart, not with the head.' The drummer was calling her again, more insistently now, and together with the 'sheer love of the work [and] the love of discovery', it was irresistible.

Algiers was becoming an increasingly fashionable choice of winter destination, rivalling the French Riviera. Arthur did not join her – Dolly had firm views about couples keeping separate interests. Before she left, and as publicity for her forthcoming book, she was photographed for *The Tatler* in a picture titled 'A Well-known Authoress', in which (unusually for her) she faces the camera full-on, wearing a flattering cloche hat and a full-length long-haired striped fur. Despite mentioning her books, the caption mostly reminds the reader whose daughter she is, whose wife she is, and (yet again) of the absence of an heir to the earldom. Such descriptions would eventually cease in the light of her achievements.

From Algiers she visited Biskra, an elegant oasis town on the northern edge of the Sahara. It was here she first felt 'the call of the southern lands that fringe it, beyond the mountains and the Thirst Countries, as I heard of them from strange men, white soldiers and black, homesick Sudanese, and dark-skinned Arabs'. As she explained simply, 'It is hard to say what makes up the thrill of exploring. The road calls and one must follow it; that is all.'[2]

To do so required proper planning – and money. For the rest of that year and during 1921 she focused on earning money through journalism with some fiction too, capturing the essence of the desert in a short story, *Where Anything Might Happen*, published in *The Sketch* in September 1920.

The young Lord Sandgate is travelling across Algeria to get over an unhappy love affair in England when he is approached by a veiled young woman, Nada, who implores him to save her from her marriage. She, too, is British and had been taken to live in Algeria years earlier by her eccentric father and, after his death, married a wealthy Arab sheep merchant. Now, after eight years, he wants a son and heir and is preparing to take a second wife, who is young and beautiful and whom Nada will have to serve. The cultural mores from her old life surface again, making her situation repellent to her, and she wants to return to England.

Enchanted by Nada and the idea of an adventure in the desert, 'where anything might happen', Sandgate agrees to help her escape, despite her warning of the dangers if they are found. He acquires camels and food, and they ride through the silence of the night to reach the safety of the border with Morocco. Above them, 'immeasurably distant, the myriad African stars glittered and danced, their white light turning the eternity of sand into a pale sea that stretched, it seemed, to the ends of the earth'. As they reach Beni Ounif, a town near the border, they stop for the night, sleeping in each other's arms. The story concludes the following morning, when Sandgate wakes to find the camels gone and Nada brutally murdered.

Dolly's story was one of a new genre dubbed 'desert romance', which was becoming popular among women readers. The trend was a very recent one, having started in 1919 when British author Edith Maud Hull, known as E.M. Hull, published her first novel *The Sheikh*. Hull had also travelled in Algeria, although only in her youth. *The Sheikh* contains passion, male dominance and abduction; two years after the novel came out, it was made into a highly successful film starring Rudolph Valentino as Sheikh Ahmed and was considered exotic and thrillingly shocking. The ending is made more acceptable for European readers by the fact that although the heroine Diana, who is falling in love with the Sheikh, thinks he is Arab by descent, it transpires that his father was British and his mother Spanish. When they died, he was raised by the old Sheikh, after whose death he returned to rule the tribe. After discovering his background, Diana confesses her love for him.

Hull wrote about the mystery and sexual allure of the Sheikh, although her heroine's love for him continued only when she found he was not Arabic by blood. She anticipated the reader's desire for a happy ending, while Dolly's stories, with echoes of her gothic-loving ancestor Horace Walpole, often had a darker take, but one that was based on her experience. With her clear-eyed view of human nature, the more she travelled the more Dolly would come to recognise the cultural challenges of sustaining romantic relationships between those of different races. While it may have been Hull's story that made the term 'sheikh' stick in the popular imagination, as Dolly pointed out, the word itself had no definite meaning and loosely meant 'the chief or head man of anywhere or anything', even just a handful of nomad tents. There were more meaningful positions of authority, she said, such as 'caid', the leader of 'a small or big town or district', or greater still, 'Agha, or infinitely greater, Bach-Aghar'.[3]

Dolly could understand the superficial attraction of the handsome, dark-eyed *sheikh*, who, 'immaculately dressed in embroidered burnous and snow-white turban, [...] struts the market place causing havoc among the more susceptible tourists', because she had seen the effect that such a figure could have. However, she also had acquaintances among educated Arab men and had spoken to wives and knew the reality of expectations of their relationships with women. Quoting an Arab writer, she noted a wife was 'A Queen by night, a beast of burden by day'. The problem was lack of education, said Dolly:

> The average modern woman, however charming her Sheikh, would scarcely remain content with the sole distractions of Arab wives, which are to foregather on the rooftops at the sunset hour, exchanging unending gossip, and occasionally go down to the 'Hammam' to take a bath.

The wives of wealthy men were more restricted as to when and where they could go out than poorer ones, whose household duties took them into the marketplace and beyond. 'The fate of the few white women I have known who have married Arabs has been heartrending,' wrote Dolly. She acknowledged there had been exceptions, 'but these few have been exceptional women or their husbands exceptionally Europeanised Arabs'.

When it was reported in 1927 that an English girl guide had married an Arab *sheikh* and had gone to live in the desert with him, the *Daily Express* quoted Dolly's gloomy prediction: 'The odds are against such a marriage being happy.' Her reasons included the undesirable consequences of her displeasing him and the ease with which he could divorce her simply by saying 'two words in front of an Arab lawyer'.[4]

In April 1920, *The Laughter of Fools* was published. Her second 'society novel', it is set during the latter part of the war when her characters are driven to extremes – the 'unhappy, hectic crowd' she had experienced.

The last year of the war and the immediate post-war period were a time of flux and difficult transition from a wartime economy and culture to one of peace, a time of heightened anxiety and upheaval for both sexes. Her characters 'drink and dope themselves blind, spend their nights in "binges", Black Masses and what are commonly called nameless orgies […] and their days in kissing and divorcing one another',[5] as one review put it. At the centre is the heroine, Louise, who has lost her parents and lives a lonely life with elderly relatives in the country until she is unexpectedly left a small legacy and finds freedom and independence, which includes two husbands and two fiancés, but in the end, she leaves for Russia to work for the Red Cross. *The Times* called it a 'clever novel […] a social "document" of an abnormal moment in the national life'. *The Evening Standard* said Dolly had 'a knowledge of men, women, books and manners; she is endowed with a gift for sparkling writing'. The *Daily Mail* considered it 'well done, with a detached scorn and detestation which save it from being merely melodrama'.[6]

She had successfully captured the zeitgeist, and everyone was talking about this fast-paced, racy book. To her irritation, however, she also discovered she had temporarily gained 'the reputation of a drug addict' because she described, 'mostly from hearsay', an opium-orgy.[7] She could not understand why people assumed that just because she was familiar with the world she wrote about, she 'must have originated all the atrocities committed by that usually misguided person'.

Just before *The Laughter of Fools* came out, so did a book by a relative. A distant cousin of Dolly's, Hugh Walpole (later sir), five years her senior, had his latest novel *The Captives* published, one of many that would see him revered as one of the best and most prolific storytellers of the time. His and Dolly's subject matter was very different, and it was not often that their names were linked, although they are bound to have met at some stage, as their fathers were close – it was Hugh's father, as Bishop of Edinburgh, who had officiated at Orford's wedding to Gladys. His eyesight having rendered him unfit for active service, Hugh worked for the Red Cross in Russia during the war, just as Dolly's heroine Louise did, suggesting some familiarity with her cousin. More often, the comparison was made between her and her late great-aunt, Lady Dorothy Nevill, from whom some said Dolly had inherited the Walpole originality.

Now the next generation was coming to public attention, with her half-sister Lady Anne making her photographic debut at the age of 5 months as the earl and countess bought Orford Lodge in Montrose and took her on her first trip to Scotland. Not that Dolly was overlooked altogether. Her father recorded that he had commissioned his wife's portrait (by Mr Keyworth Raine) for £300 that year, and noted, 'I also bought a picture of my daughter Dorothy painted by Mr C.B. Prescott. I gave for it £84.'[8]

The publication of *The Laughter of Fools* seems to have secured Dolly's status as a novelist, which description often now accompanied her appearance in the press. She was also becoming the 'go-to' writer on the sexes, with her bons mots quoted. The *Pall Mall Gazette* devoted a section to her witty phrases from *The Laughter of Fools*, such as 'In her own world, the women seemed to consider love as a means to marriage, and men to consider marriage the means to love',[9] and 'A man will tolerate a good deal of criticism but he won't tolerate aspersions on his lovemaking'.[10] In one section of a newspaper called 'We take off our hat to …', which was dedicated to well-known figures, her picture appears next to that of Winston Churchill and she is quoted as saying, 'The only perfect husbands are other people's'.[11]

She may well have been seeing more of those than her own – Arthur only occasionally accompanied her to social events, which was as she preferred it, since she thought couples should have separate as well as joint friends. He was also busy with his own work and there was talk of them collaborating on a novel together, although nothing was published in joint names. However, they did get into the habit of putting work aside and holidaying together for part of the year. In 1920 they went to the USA and Berlin to see Arthur's army friends who were stationed in post-war Germany.

By mid-1921 they were living in Ebury Street in Belgravia, a very desirable part of London, albeit in a modestly sized apartment with a housekeeper. The location was convenient for socialising, in Dolly's case, to indulge in her favourite pastime at the newly opened dance club, The Diplomats in Bond Street, a large, lavishly furnished venue with a glass roof to keep its clients cool.

Her journalistic reputation was enhanced by a series of articles on men, women and love commissioned by *The Ladies Field*, an upmarket weekly paper, and amusingly illustrated by Scottish artist Alistair K.

Macdonald. The series was publicised in advance by a full-page photograph of Dolly taken by the sought-after photographer Yevonde, with a fulsome caption referring to her novels, which were written 'with merciless truth of her own generation'. Comparisons were made with her contemporary, Stephen McKenna; unlike him, however, she distinguished 'between symptoms and the disease'.[12]

With her usual perspicacity, her articles analysed, not without humour, the emotional differences between the sexes and their attitude towards relationships, especially marriage. Although the number of divorces had increased since the war, it was expensive and still carried a stigma. Further, while men could rely on their wife's adultery as the sole ground for presenting a petition, women had to prove an additional ground, such as cruelty. '[I]f marriage for a woman proves a failure,' Dolly wrote, 'it is far, far harder for her to disentangle herself from it, and the pains and penalties for her are so much the greater.'[13]

Her views were frank, her approach direct. Much of the problem for women, said Dolly, in her first feature for the series called, 'What is Wrong with the Woman's Point of View on Love?', was the imbalance of power. Having always relied on a man for her protection and that of her children, a woman knows 'what it is to fear man's disfavour, or the cooling or straying of his affection'.[14] That makes a woman too cloying, too anxious to be the object of his entire existence, becoming 'jealous of his work or play or sport, and most of all of his friendships'.

A woman was too inclined to treat love as a 'cataclysmic or supernatural' phenomenon, rather than a natural one, said Dolly, which was partly the fault of her upbringing:

… her half-baked education, the exaggerated sex-consciousness mixed with ignorance that has been inculcated into her early girlhood, and the factitiousness of the books she reads and the conversations she hears. She has learnt to regard love out of all true relationship to life.

This could affect the way she behaved in a relationship: essentially, she needed to let go a little.

At the same time, Dolly did not believe there was 'any great difference between man and woman in the great essentials of love […] I think at heart they are much the same, and have the same needs, emotions and passions.'[15] However, women had 'an infinitely more complicated and subtle range of emotions' than men, which was where men's problems arose. 'This is where calm acceptance loses him the woman he loves.' A man believes that where there is mutual love and kindness, 'things will flow serenely and indefinitely without further effort on his part. [H]e takes it for granted that so long as a woman does not show any active signs of grief or anger, she is as serenely content as he is himself.' As a result, he upsets her in many ways without realising it.

Dolly's overall view of men was critical. She liked them but considered them flawed, although not irredeemably. In a feature called 'Are Englishmen Jealous of Foreigners?', she wrote of their suspicions when it came to women. 'I believe the average Englishman would prefer ten co-respondents [men named in a divorce] of British extraction to one platonic foreigner having tea with his wife'.[16]

The reason was simple: 'After many centuries of a somewhat feudal system of life, [the Englishman] has learnt to believe [...] that he is invincible. He has athletic prowess, a good seat on a horse, undeniable good looks and is "plus" at golf; what more can a woman want?' He tends to take a faithful wife for granted; with women in general, he does not feel it necessary to exhibit 'charm of manner and social polish' in case he might be thought 'a ladies' man' or even 'effeminate'.

The foreigner, on the other hand, takes great trouble with women. A woman is interesting to him *as* a woman, 'not necessarily as a flirtation or hoped-for *affaire*. [...] Each woman with whom he is temporarily concerned is a subject of special study; he takes infinite pains to please her and to keep her pleased.' By contrast, the Englishman often behaves like 'an exultant cockbird' that struts about trying to impress its mate, not thinking that she may have her own preferences in how she is wooed. When he sees the foreigner spending time talking to a woman about subjects that interest her and trying to understand her feelings, the Englishman resents and fears it.

Dolly saw Englishmen as lazy and self-absorbed in matters of love. If an Englishman is turned down by a woman he tells himself she has no heart. 'A cleverer man under the same circumstances says, "What have I done wrong?"' The remedy for the Englishman's unease and jealousy lies in his own hands, because (she said, loyally) they are the best-looking and best-dressed men in the world, and 'except when they try to be, they are not really stupid. But,' – and here is her crucial point – 'this is a day of mind over matter, especially in the advancing mentality and freedom of women.' Modern woman would rather be amused than impressed and 'would rather be interested than kissed. She has outgrown her content-ment with a splendid figurehead, as the intelligent nursemaid has with her infatuation for a scarlet tunic.' If the Englishman took a fraction of the trouble with women as the foreigner does or that he himself takes over his sport, he would have no fear of rivalry.

Little wonder that in her feature called 'Does Man or Woman Suffer Most in Marriage?'[17] she firmly concluded that woman does, using arguments and examples that are (regrettably) still relevant a century later. Whether she stays at home or goes to work, said Dolly, the woman ends up doing most of the domestic work and organisation, 'And in the intervals she must be the Eternal Feminine, The Queen of Hearts and the Patient Griselda!' She concluded that the pledge 'for better and for worse' means, ironically, that 'In view of the shorter youth and longer faithfulness of women, the permanency of marriage is to the average woman very often its saving grace'. Her feature 'The Advantages of Being a Plain Woman' was surely influenced by the remark made about her own appearance. While she acknowledged that 'every normal man is attracted by a pretty face', she considered that modern man needed more from his wife – he wants more than 'a doll to dress up or an *objet d'art* for the adornment of his house'. (The term 'trophy wife' had yet to be coined, the need for which would surely have disappointed her.) That is where the plain woman has the advantage, for she makes the most of her personality and other assets and sets up 'a store of interests, occupation, sympathies which last her long after the unkind years have reduced her and her lovely sister to the same dead-end of pulchritude'.[18]

Little in modern manners escaped her. In her feature, 'Why do People Laugh at Nothing?', she mused on why there seemed to be 'a fashion for laughter that often scarcely pretends to be real [...] It cannot be for any aesthetic reason. Laughter can be one of the most attractive things about a person but it so rarely is!'[19] because it produced much stretching and straining of neck muscles and 'unattractive sounds'. She put it down to lack of assurance: people feel they may not be sufficiently strong in their powers of entertainment or expression, so they draw attention to themselves by making 'a noisy outward display'. By contrast, when there is a gathering 'of real wit and intelligence', there is only an 'appreciative nod or quiet chuckle'.

Since the war, Britain's old class structure had become less certain and more fluid, and on this, Dolly quoted her great-aunt. One of Lady Dorothy Nevill's later books talked about laughter. To her, it seemed that those who habitually indulge in it 'are not quite sure that they are ladies and gentlemen and are therefore perpetually trying to laugh it off'.

Not only was Dolly producing features, but she was also completing her next novel, *The Tent of Blue*, squeezing in time for relaxation by going to the theatre or concerts and mixing with the new Russian émigrés who had escaped the revolution and were finding their niche among London's artistic crowd. Arthur, too, had finished his second novel, *Pillars of Salt*, published, like his wife's, by Duckworth & Co. and advertised alongside hers. Its subject of post-war society was similar to Dolly's, containing hunting, gambling, drugs and adultery. The heroine, Ursula, had appeared in Arthur's first novel, *Ursula Vanet*, published the previous year, and was described as a 'married flapper'. The term 'flapper' had been coined post-war to describe women and girls who were too young to vote (therefore, under 30) and was often used to indicate irresponsibility, even immorality; sometimes, however, it simply meant a modern woman. While reviewers agreed that *Pillars of Salt* was a good-humoured book, some damned with faint praise. 'It has no literary pretensions but is made readable by its air of cheery realism,' said *The Daily News*. 'It has neither sentiment, subtlety nor passion [...] Quite passable superficial comedy.'[20]

Dolly's *The Tent of Blue* recreates the atmospheric beauty of the desert. In it, an unhappily married woman with an excessively jealous husband goes to find healing in Algeria. There, she falls for a British explorer, endures danger and bloodshed with him and, torn between pity for her husband and passion for her lover, she returns to London with him, with tragic results. (This time, Dolly found she was credited 'with keeping lovers in the Sahara'.[21]) Reviews were generally very good. One called it 'a Greek tragedy brought west and modernised'.[22] *The Spectator* enjoyed the vividness of her desert descriptions. *The Times* also liked it, referring to the pathos of her heroine, and said Dolly 'knows the value of restraint and does not paint in her local colour too heavily'.[23]

The same person reviewed a novel called *The Jewel in the Lotus*, also set in the desert, by another writer and traveller, Rosita Forbes, but found it wanting. Coincidentally, the two women knew each other – Forbes' brother had recently married Dolly's cousin, Maude Walpole, and Dolly had been a guest at Forbes' (second) wedding.[24] Both women would later be lauded for their different exploratory achievements. For now, though, Forbes' novel had disappointed the *Times* reviewer, who had hoped she would 'make use of her own unique experiences as traveller and explorer to give life to the fictitious life of her heroine. We find that this is so, but not to the extent or in the spirit we had hoped.'[25] The reviewer found Dolly's heroine 'more appealing' than Forbes' 'because there is less pretentiousness about her'.

Elsewhere a sardonic feature mused on whether authors wrote for pleasure or money, and in giving a summary of *The Tent of Blue*, said, 'Whether Lady Dorothy Mills writes for business or pleasure I know not, but one always expects people of title to be very wealthy, so perhaps she does it for fun.'[26] No doubt Dolly laughed bitterly.

Meanwhile, she had been preparing for her first expedition, to the reclusive cave dwellers of the Tripolitan Mountains in Tunisia, also known as troglodytes. Tunisia was then a French protectorate (remaining so until its independence in 1956). She had first heard about the cave dwellers when she was in the South of France, a few years earlier. Visiting the little town of Chissay, where some of the people lived in ancient caves carved out of the overhanging rock (albeit with modern comforts), she was told they were the descendants of an ancient desert tribe, their ancestors having travelled across to the Mediterranean centuries earlier. Dolly's explanation of what spurred her desire to find out more about the cave dwellers was recounted by *The New York Herald* in what may have been a rather fanciful feature.[27]

A young woman in Chissay, who had been adopted by a local family, apparently told Dolly that she had been brought there as a child, wearing desert clothing, by a strange being whose language was unintelligible but who bore a resemblance to the local people; the young woman believed he was part of a lost desert tribe to which she too belonged. She had vague memories of living outdoors, sandstorms and a boy playmate but nothing more. While Dolly was sceptical of how she could have been saved by someone allegedly from the desert who knew exactly how to find her 'people' in Chissay, the idea of a lost tribe was appealing, and she had determined to find out more one day. A prince in Tunisia had promised to give her men to guide her and a camel to carry her. He told her the tribe lived by an oasis 'a day's ride from the elephants' graveyard', and although no one knew exactly where this was – such places are potentially mythical – Dolly remained determined to find them.

Hyperbole is nothing new, but some newspapers relied on the fact that most readers did not have the opportunity or means to travel abroad, let alone to Africa, and so they were eager to exploit the myths and fears evoked by what was then often referred to as 'the Dark Continent'.[28] Dolly would soon find their approach challenging: she wanted to give the readers an accurate picture of what she experienced, making it exciting but not unreal (notwithstanding the occasional use of dramatic licence). However, she would soon find that sensationalism was often required – and if she did not provide it, the newspapers would. A comparison of certain experiences described in press features about her travels with her description of the same events in her books usually shows the latter to be more controlled.

If a feature was not written by Dolly herself, the best source of information came from interviews she gave, although she rarely included much detail on the practicalities of arranging her expeditions: no doubt she considered it was the least interesting part of her story. While the first modern British passport had very recently come into being, other permissions were usually necessary, as was the employment of local guides for navigation, porterage and translation. Even when her name was sufficient to get the attention of the relevant authority, permission to travel was not guaranteed and when it was forthcoming, it was not always granted graciously.

Her Tunisian expedition took place in the spring of 1922, when the average temperature was comfortable, similar to that of an English summer. Travelling 150 miles inland, Dolly was the only white woman known to have stayed with the cave dwellers and experienced the lives of their ancestors in caves designed for coolness and protection.[29]

Her journey began at Tunis, from where she went by car to Gabos and on to Tataouine,[30] one of the last military forts in North Africa, where she was able to engage a group of Mokhazni horsemen. They escorted Dolly, now also on horseback, to the cave dwellers, who were Berbers descended from Stone Age tribes of North Africa. Their caves were cut into the solid rock 'situated on a range of hills covered only by tufts of alfa and coarse scrub'.[31] She visited several such villages, including Chenini, and described her experience in one of them.

On her arrival, the whole village turned out to greet her; it was deceptive how many people the caves housed. As she was wearing riding jodhpurs, they could not understand how, if she was female, she could be dressed in such a way and they referred to her as 'the one with the face of a woman and the legs of a man'.

The people would have been informed of her visit in advance, both out of courtesy and to minimise the chance of any hostility. They had often been suspected of devil worship, but Dolly said they were in fact 'of the Mohammedan religion' and had more recently carved a mosque of sorts out of the rock.

After enjoying a banquet given in her honour, she retired to her cave:

A narrow chamber, very stuffy and totally bare of furniture. There was no bed but, following the custom of the men and women, I rolled myself in a camel rug each night and the cave men, who were determined no harm should come to their white woman guest, built up the doorway with wood.

As she would often find, she aroused more curiosity in the women than the men and she was followed everywhere. 'They gazed at me in awe and wonder and took the greatest interest in everything I did. They came into my cave and watched me at my toilet.' The women had straight black hair, which they groomed with wooden combs, and they could not understand how Dolly's hair, pinned up for comfort, was curled. Her skin, so different from theirs, fascinated them and they would gently stroke it. She enviously noted their jewellery of beaten silver and gold and watched them elaborately make up their eyes using 'long sticks dabbed with a black preparation'. They made up Dolly, although she found it a painful process. In return, she let them use her face powder, the like of which they appeared never to have seen before.

Their relationship with the men held particular interest for her. Courting rituals barely existed. The usual age for marriage was between 14 and 15 and if a man saw a pretty girl:

> He simply goes to her people and arranges to provide her with a trousseau. She comes to him with nothing and his sole duty is to give her an outfit. This may consist of camel's hair, cloth or silk, which has been got in bartering with the Arabs of the desert.

If a girl reached 18, however, her chances of getting a trousseau were reduced; by 20, she was regarded as middle-aged. Sometimes, if a man heard of a pretty girl in a neighbouring tribe, he took a chance without even seeing her. Dolly noted wryly that at least divorce was easier than in England: the day after the wedding, if he decided he did not like her, he simply told the *sheikh*, took back what he had given her and returned her to her people, providing her with a donkey to carry her if necessary.

For certain sections of the press, the fact that these people were cave dwellers gave rise to an irresistible headline. 'How the Caveman Makes Love' headed a feature in *The Washington Times* written after an interview with Dolly,[32] which must have exasperated her, although she did not mind sharing elsewhere a proposal she had received from a *sheikh*. Sending a message through an interpreter, he said he liked the look of her and did not have a white wife 'and, if I would marry him he would make me his favourite. Of course, they all say that, don't they? [...] I sent him a kind reply, telling him how honoured I was by his offer and explaining that I was already married.'[33]

As for her status, the women asked how many wives her husband had – their men were allowed four, although generally took only one – and when she said she was Arthur's only one, they could not understand why she was away from his side. In the week she lived with them, Dolly tried to tell them about Britain and how she had travelled but they had no concept of the sea. 'They are a peaceful people,' she said, 'who, to my surprise, had never heard of the great European war and knew very little about the ways of civilisation.' Yet they had medical skills: she was told they could trepan a broken skull and, using boiling tar in place of anaesthetics, perform successful amputations.

Dolly's interest in religion and ritual and the relationship with magic would infuse her writing and fascinate a wide readership. In the Holy City of Kairouan in central Tunisia she was confounded by what she witnessed among the Aissaouia, an ancient Islamic mystical brotherhood which practised complex religious rites that to Dolly seemed 'more demoniac than divine'. Boys and girls of 4 or 5 years old 'ran to and fro plunging long metal skewers through their plump cheeks and arms and legs'.[34] A man, howling like a wolf, knelt while the imam used a heavy mallet to drive a 3ft sword through his stomach. As the sword was withdrawn, the man seized it 'with an expression of horrified rapture' and drove it in again while he stood, 'howling and swaying in a bear-like, rhythmic dance [...] When he withdrew the sword, there was no trace of blood.' The fuller description of the rites she gives in her book is even more graphic.

Dolly was interviewed after her return for *Motor Owner* magazine, in which she recommended parts of Tunisia to keen drivers and other tourists. The respected journalist Clive Holland noted, as others would do, 'It was a little difficult for us to imagine that the speaker – slight, pretty, and almost frail looking, but vivacious and evidently keen – could have travelled where she has, and alone.'[35] While her appearance belied her strength and sheer grit, Holland had hit on a point that distinguished Dolly from many other travellers in that she always chose to journey without a companion, preferring to rely on local guides and carriers.

Her impressions of the country lingered. 'There is a great deal in the fatalism of the desert people,' she wrote, 'and that is why, although you may hanker after the fleshpots and return to civilisation for a time, the desert always claims you again.' Four months later, she left from Liverpool on the expedition that would define her both as explorer and acclaimed travel writer.

WHERE ALL ROADS LEAD

Dolly was never one to shirk her commitments and while preparing for a major expedition to Timbuktu, she was also busy on the committee for the Three Arts Ball. Held every December at the Royal Albert Hall, it was essentially a huge and extravagant fancy dress event for members of the Chelsea Arts Club, which became increasingly flamboyant and scandalous as the years went on. At around this time, there was a set emerging, dubbed by the press 'the Bright Young People'. They were mostly aristocratic, and in reaction to the horrors of war and the fact that for many of them their material situation had changed for the worse, they developed their own code of social conduct that reacted against the pre-war expectations of their class. They delighted in decadence and flaunting their sexuality, and even though homosexuality was illegal, there was generally an indulgence towards it.

Costume both disguised and revealed. On this occasion, however, Dolly's was a modest but interesting Arabic travelling outfit, which gave her the opportunity to talk about her forthcoming journey and saw her on the cover of *The Tatler*.[1]

Soon she was away to live a very different life, at least for a few months. Dolly called herself an explorer, as the world would do, but she was careful to differentiate between the title as used at that time and its original meaning, which was 'a person who travelled in, or actually discovered, a country or tract of land hitherto unknown by experience or even merely by hearsay to civilised man'. In fact, she thought it doubtful 'if such a place of any magnitude still exists'. Instead, the words 'explorer' and 'exploration' were being used more freely 'in connection with the travels, scientific or merely spectacular, of a great many people'.[2] She therefore considered herself an explorer as defined by the dictionary of the time: 'To search by any means; to enquire into with care.' For her, exploration was 'more than the gratification of the Wanderlust [...] it is the spirit of the collector'.[3]

However, no collection, whether of '*objects d'art*, or love affairs, or race horses', was more gratifying or thrilling 'than the discovery of an unknown tribe, a new oasis or the right to trace a new name on a map'. There certainly had to be a strong reason to impel a person to leave behind all home comforts, friends and family, and exchange them for 'uncomfortably and dubiously safe places, to endure seasickness, fatigue, heat, cold, and privation and fear [...] to eat mysterious horrors [...] and to smell all manner of unexpurgated smells'.

On 30 December 1922, a week before Arthur left for Argentina to gather material for his next work, Dolly sailed for West Africa. 'One has no idea of the size of West Africa until one goes there,' she wrote. 'It is really quite a long way to Timbuktu but it is only about three inches on the average map.'[4]

The timing of her journey was important. Just weeks earlier, the British Consul-General for French West Africa had warned that Timbuktu and the Great Bend of the Niger River could only be reached with any certainty during late November and early December; much of her journey would be by water, so she was cutting it fine. 'I hate all the obvious stereotyped routes,' she said in an interview. 'I love the wilds, new places and curious things, and I hope to pick up copy that I may embody later on in a novel based upon my travels to Timbuktu.'[5]

But there was more to her choice than that. She had first become aware of Timbuktu as a child when she read a piece of nonsense verse by Samuel Wilberforce, an Anglican bishop:

If I were a cassowary
    On the plains of Timbuktu,
I would eat the missionary.
    Bible, prayer-book, hymn-buk, too.

Situated in Mali, Timbuktu lies on the southern edge of the Sahara, a few miles north of the Niger River. For centuries, it was a major commercial and intellectual centre but after being captured by Morocco in 1591 it declined and was repeatedly attacked and conquered by African tribes. In Dolly's time it was under French rule, having been captured by France in 1894. Although they had partly repaired the city from the desolate state in which they found it, no connecting railway or hard-surface road was built. Getting there posed something of a challenge.

As she would explain in her first travel book, *The Road to Timbuktu*, an absorbing blend of adventure, observation and history, while she was in the northern Sahara she had looked down 'the great invisible roads' leading south, all of which seemed to lead to Timbuktu or another city in the southern desert, 'and I longed to travel those roads, those fiery roads that lead into the arms of the sun'.[6] The mystique of the ancient city intrigued her:

All the old caravan routes go there; the great cruel roads of the ages, along whose blazing trails, century after century, have toiled long trains of men and animals, bearing gold and spices, ivory and cotton [...] roads on which men have died in their thousands [...] to gratify the whims of long-dead Emperors and strange Queens.

Timbuktu had been a centre of the slave trade too, where women 'were added to the precious merchandise to serve as foils to golden-haired Circassians in the palaces of the north'.

Europeans had seldom fared well there. In 1828, less than a century before Dolly's journey, a French explorer, René Caillié, was the first European to return alive, largely because he had disguised himself as an Arab and invented a persona. His predecessor, Major Gordon Laing, a British Army officer, had been murdered there two years earlier as he was leaving.

Dolly knew this from tales she had heard about:

... the old days before the French occupation filched most of the romance from the Sahara; [tales] of guerrilla warfare and the taking of villages, of curious tribes and customs, of the sluggish fever-laden rivers of the south [...] of the predatory Touaregs [...] of utter desolation and thirst.

She learnt, too, about the men who had made French colonial history, 'whose bones, bleached and polished, have made landmarks for passing caravans'.

Within the lyricism of Dolly's prose is frequently found a personal, often humorous touch in the relating of her experiences and feelings which takes her writing beyond the mere narrative. After leaving Liverpool, 'on a wet nightmare of an afternoon in late December', SS *Prahsu* took her away from the greyness of the River Mersey and eventually the days grew longer and warmer as they slipped down the African coast, past the Canaries, looking in at Las Palmas and Tenerife, 'whose Casino has reason to rise up and call me blessed', and just as the snowdrops were starting to appear in England, on the bluest and hottest afternoon of all, they arrived at Dakar in Senegal, seat of the French colonial government.

One of the new experiences facing Dolly who, so far, had met only Arabs, would be travelling and living among Africans. She was looking forward to it and, as a keen boxing fan who had reported on a major fight in which the French Georges Carpentier became world light-heavyweight champion, she was pleased to be in Senegal, birthplace of another boxer known as Battling Siki. 'I just missed his fight with Carpentier in Paris, worse luck,' Dolly told the *Liverpool Echo* before she left, in which Siki had proudly taken the title.[7]

Africa was fundamental to understanding the 'race problem', which was increasingly seen as the major issue of the twentieth century, at a time when racist representations still pervaded British culture. During the interwar period there evolved a more liberal ideal of progressive and enlightened imperialism premised on improving 'race relations' and thereby forestalling the development of a more radical anti-imperialism that could threaten the whole fabric of the empire. It became fashionable for liberally educated middle-class British women to take up worthy political causes and generate new debates around race, class and imperialism, a factor in the development of race consciousness at a time when the black population was increasing in Britain.

Dolly's travels were mostly in non-British colonies, and she did not agitate for racial equality as did other prominent women, such as Nancy Cunard and Winifred Holtby. However, despite championing black rights and famously taking a black lover (who was American), Cunard never went to Africa, unlike Dolly, whose writing about her experiences would be of great topical interest.

Neither would Dolly have any truck with clichéd representations of Africa. Many times, when she was in the African bush, she 'wished for Mr Charles Cochran', who was the most famous theatrical manager and impresario of the time, producing plays, spectacular musicals and revues. Her wish was not as strange as it may seem. Whenever she saw 'something that is more beautiful than anything I have seen on the European stage', she wanted to share it with him. 'Only his intuitive imagination could catch the special beauty and character of primitive Africa, its utter unlikeness to anything else, especially to the Africa usually represented on the stage – a mixture of Livingstone and Barnum's [circus]!'[8]

From the outset she had an awareness of issues surrounding colonialism and did not travel with an uncritical eye. As she would write in the preface to *The Road to Timbuktu*, when travelling in colonies of countries other than Britain, she found:

> … much that was new and interesting to me in their problems, and in their manner of solving those problems. But though I found very much to admire and even to love, I do not feel qualified to generalise, to make comparisons, or to air my personal opinions in print.

However, she often found it hard not to. When it came to the topical and controversial question of whether to 'civilise or not to civilise', her conclusion was that generally white influence was not a good thing. She also preferred the term 'undevelopment' to others in use that suggested backwardness.

At Dakar, landing among mountains of peanuts awaiting shipment, she picked up her guide and a driver as arranged and they continued into Mauritania, the land 'forgotten by Allah', where they hit a sandstorm: Dolly's first, it was a devilish experience. On they went to Bamako on the Niger, where the French authorities told her it was foolish to try to go further, of which she took no notice. At Kouakourou they ran into the worst of the mosquito region, fiends 'that no net could keep at bay, that made the hours round sunset and the nights an agony of scratching, till the blood ran from every inch of my body'. She had brought various insecticides from England, but 'they simply gorged on them and asked for more'.

She learnt to keep an eye on her guides, not just for her own sake but other people's. Her driver, Saghair, was usually quiet and early to bed, but at Kouakourou he disappeared one evening into the village and did not reappear until the next morning, with Dolly's breakfast (of bitter coffee and dry bread). 'Ten minutes later appeared a brace of angry parents,' recalled Dolly, 'followed in the distance by a sheepish-looking girl. This latter was Saghair's great sin, and the parents, incoherently wrathful, expatiating on her previous purity of driven snow, seemed to think I had something to do with it.' Feeling slightly apologetic, she offered a goat and a handful of kola nuts, 'which, judging from their triumphant mien and profuse thanks, was considered a generous price for a Soudanese maiden's honour'. As they left, she noticed the girl look coyly back at Saghair, leading Dolly to warn him that his behaviour could not continue, on financial or moral grounds, and that if it did, the goat and the kola nuts would come out of his wages. However, 'Saghair, like Adam in the garden, shifted all responsibility: "*Elle pas bon. Elle m'a cherché*," he said.'[9]

No news reached her from or about England in her first weeks away, and to her shame, she found she scarcely cared whether it still existed. 'It was a curious effect that the big spaces have on one, seeming to absorb one utterly.' She had been homesick and lonely in big cities, even when surrounded by people she knew, but never in these big spaces. 'All the everyday problems and interests, little joys and little sorrows of civilised life, seem trivial and superfluous [...] One becomes just a healthy animal.'

At Mopti, the last outpost of any size in Mali, she saw a white child, about 2 years old, the only one she ever saw there, and realised that 'someone's wife, greatly daring, had borne it into that burning fiery furnace in which, contrary to accepted ideas, it seemed to survive a good deal better than its elders'. Mopti marked the end of overland transport and from there she had to journey by river, her vessel now a simple river cargo boat that she dubbed 'Tin Lizzie'.

In villages beyond Mopti lived men with teeth 'filed to long, sharp points', who were related to cannibals, and it was here that she had an unnerving experience which taught her something about herself. She was strolling through the bush near a small village on her own when an 'enormously tall' man with pointed teeth and wearing fearsome-looking ornaments suddenly appeared and started walking with her. She said, 'Good evening' in her best Bambara (the national language of Mali) and when he did not answer, tried Arabic and snatches of other dialects but nothing elicited a response or a smile. Dolly wanted to return to the village but was not sure of the path she was on; she did not wish to appear vulnerable, so she kept going. Not only did he walk just out of her eyeline, behind her shoulder, but a shape under his clothing suggested a knife.

They had walked for nearly an hour, the sun getting hotter, sweat pouring down her face, 'and I longed for a sight of that smelly mud village as I have never longed for London, Paris or New York'. She told herself that the district was safe:

> But every time, out of the tail of my eye, I caught a glimpse of that silent, stealthy, stalking figure, with its hand to its side, its pointed teeth and staring eyes, I shivered […] and the pad-pad of his bare feet seemed to chant all the cannibal stories I had ever heard or read.

Relief replaced fear when she realised he had disappeared into the bushes – but then she heard the familiar pad-pad of his feet and realised he was running towards her. 'I only hoped […] he would get it over quickly, that I should meet death as an Englishwoman should, and that a few people would be sorry and that somebody should pay my bills.' As she turned to face him, he stopped and held out his right hand to reveal a big bunch of *shum-shum* berries which, with a big grin revealing his pointed teeth, he thrust into her 'nerveless' hand and disappeared into the bush as silently as he had come. As the reality of the situation dawned on her, she sat down in the dust and laughed. Taking the berries home for her supper, she was annoyed with herself on two counts, 'I – who had always flattered myself that as a woman and a novelist, I could recognise a man's intentions – had lost a most wonderful opportunity of taking some photographs'.

As they neared Timbuktu, chugging along the river at 4mph, she mused on the perversity of human nature: 'It is a depressing fact that so many of the long-anticipated moments of life are obscured by the sordid insistence of one's vile body.' For now, it was what to eat and how to get it. She was sick of stringy chicken and tinned food, and bread when she could get it that had to be soaked in river water for twelve hours to make it chewable. She gave thanks to the inventor of Worcester Sauce because it alone 'disguised the various stalenesses and the ten-days-old eggs that got eggier every day'.

One day, Saghair found a nest of crocodile eggs and Dolly cooked two of them and thought they were quite palatable but was violently sick that night. She was well aware of the danger of the huge beasts in the rivers, having been warned of regular instances of human killings, to the extent that she considered crocodiles to be a far greater threat to African lives than any leopard, lion or even cobra. But the enemy dwelt inside, too: the many creatures that crawled and bit her in her frequently collapsing bed on the bucketing boat saw her crave a decent night's sleep.

At last they reached Kabara, the ancient commercial port of Timbuktu, where she hoped horses had been arranged to take them the last few kilometres to the city, a journey of roughly thirty minutes, but instead, she found her transport was disappointingly a barge. For four hours, through monotonous scenery with the sun beating down 'pitilessly, mercilessly', they moved along narrow channels, the men punting so slowly with their bamboo poles that Dolly thought they would never get there. Fortunately, halfway along, a group of officials sent by the Commandant of Timbuktu turned up to take over and propelled the barge along at rattling speed 'with rhythmic stamping of feet and a hoarse chanting to which their feet and their poles kept time'.

Suddenly, a narrow canal 200 yards long lay ahead and at the end of it, silhouetted against the sky, stood Timbuktu. It was not, however, the breathtaking experience she had anticipated. 'For days and weeks before I reached Timbuktu I had dreamt of the moment when I should first lay my eyes on that legendary, mysterious city,' she wrote. 'When that moment came, my eyes were so shut up with mosquito bites that I couldn't have told it apart from Piccadilly Circus.'[10]

What she could see looked promising at a distance, the crumbling grey walls lying 'like some animal, aged yet imbued with dormant fierceness that, crouching, waits to spring'. However, her second impression was not inspiring, rather like that of René Caillié. Nothing had met that explorer's expectations either. He had seen only 'a mass of ill-looking houses built of earth. Nothing was to be seen in all directions but immense plains of quicksand of a yellowish grey colour.'[11] Dolly would add to his impressions 'the vulgarising touch of modernity'.

Still, she was greeted with charming cordiality by the commandant, a smart military figure in spotless white, who took her to lunch in his house, en route to which she noted the ugly grey military buildings and a large square with a 'horrid' attempt at a garden in the middle of it. Beyond all this, though, she was thrilled by a glimpse of 'a labyrinth of grey streets, tortuous and crumbling [...] with a suggestion of mystery, a hint of reticence, that one only finds in the towns of the Sahara – the mystery, the reticence of Islam'.

Comfortable accommodation being scarce, she stayed at first at the commandant's house, a large, galleried building where she relished having two rooms, high and cool, 'with a real bed and a real washstand'. She wandered outside, finding that the sand came up to her ankles and became aware of the faint and familiar smell of burning wood and incense. And then, 'with diabolical perversity', it all went downhill. After leaving 'home and duties and pleasures' to travel 3,500 miles to the place she had dreamed of seeing, she fell prey to an attack of what she said was influenza that saw her bedbound for a week, made even worse by a heatwave – 'and a heat wave at Timbuktu is not an empty phrase! [...] Everyone was kindness itself, except Saghair, who [...] seized the opportunity to go on a prolonged "bust" round the town of his forefathers, leaving me helpless and unnursed.'

All she had seen of the city was a quick walk around its famous marketplace, but being bedbound gave her the chance to become attuned to the place. All the sounds of life drifted to her: from 5.30 a.m., the crowing of deep-throated cocks; from 6 and at every half an hour, the military bugles; the voices, shrill and guttural, from the meat market;

and always, 'surely the most characteristic sound of West Africa', that of corn, or anything else that needed to be pounded in a big wooden bowl with a wooden pestle, whose 'rhythm and tempo became as familiar to me as a well-known piece of music'.

As a heavy hush fell over the city each midday and the heat and glare bore down on the flat roofs she could see from her open door, she imagined in her fever a house or minaret towering up in the air, or the whole of the city shape-shifting into a mirage as seen by men dying of thirst in the desert. By four in the afternoon, the sounds changed again, '[T]he voices of humanity rose, polyglot and guttural – the shrill giggle of women, the crying of babies, the deep-throated murmur of men, and a low hum reached me even from the big market place nearly a kilometre away', followed at sunset by the prayer of the muezzin from the mosque.

At night, still disobeying the military doctor's order to keep her doors and windows closed, she would hear Arab music that took her wandering through the oases of southern Algeria, and finally, a new sound, 'a faint, sad little air played on a *balafou*', which was restful and unmistakably African and was the music that lulled her to sleep.

When she was able to sit up and eat without being spoon-fed (for which Saghair returned to carry out his duties), 'a bit of the desert would walk right into my room – a shaggy Touareg, with his veiled mouth and loose blue draperies [...] carrying effortlessly on his head a huge goatskin of water, from which he filled the great earthenware jar in the corner of my room'. Other visitors would drift in and out without knocking, Berbers from the north, selling leather bags and ostrich feathers, who spoke gently and sympathetically and never seemed to mind if she did not feel up to examining their goods. A beautiful woman popped in too, curious to see this visitor and, keen to chat about feminine matters, proudly shared with Dolly the name of the upper-class white man whose mistress she currently was, and the army general whose mistress she had been the previous year.

It was women who psychologically interested Dolly more than men, 'for man may show one the mind of his race, but women show one its heart'.[12] Particularly in the regions in which she travelled, she felt women were more likely to confide in their own sex than in a male traveller: '[A]nother woman of no matter what size, shape or colour is to her way of thinking a sister, an equal, an ally in the great war of the sexes and to her she will confide, to the smallest encouragement, her secrets, her aspirations and her problems.' Whatever their colour, women remained the same underneath, said Dolly. 'It is only the man-made standard that a man has set up that changes and that in each country woman copes with or counters for her own ends.'

Never intellectually idle, as she recovered Dolly learnt more about the history of Timbuktu from old books and papers brought to her, which she would relate in her own book. She read of its vicissitudes of fortune as a byword throughout Africa for brilliance, learning and prosperity in the late sixteenth century; its conquest by the Moors, 'one long story of bloodshed and pillage and looting'; and its period of 'sleep' from 1730 until the arrival of the French in 1894: they were the first white men to take possession, which triggered a massacre by the Tuaregs and bloody revenge by the French. Dolly would surely be saddened, if not surprised, to know that a century after her visit, the area remains troubled.

She read of the ancient history of the region's trading, especially with the Mediterranean countries, which included providing slaves and gold dust in exchange for desirable products. Timbuktu was also a city of pleasure. An old chronicler described the moral standards of the area between it and another city, Gao, as 'The extreme limit of immorality. The gravest crimes, the most disagreeable acts, are committed openly, and the worst turpitudes are spread out in the light of day.'[13] Of the last days of the Sonhrais Empire, a seventeenth-century chronicler said they drank wine and 'gave themselves up to unnatural sin', as well as practising adultery so frequently 'that without it, there was no elegance, no glory,'[14] about which Dolly noted drily, 'In fact, it all sounded rather like the best London society of today!'

Dolly's curiosity about the people she met throughout her travels was rather like than of an anthropologist but with a more personal approach and enhanced by the skill of being able to communicate her observations to the general reader. In all her travel books, she excels in describing and distinguishing between the cultures she encounters, for which Timbuktu gave her good practice with its melting pot of races that made up its troubled tapestry. Here are the nomadic Tuaregs, a tall, handsome, warrior-like people who were always veiled; the Sonhrais (or Songhai), whose empire for centuries had been the largest in West Africa and who were the resident race in Timbuktu; their relations, the Rimaibé, the descendants of slaves; and the Bozos, whose women, like the Sonhrais, wore their heads 'shaved all over, with the exception of a number of little round tufts of hair sticking up like balls of wool'.

Here too were the Moors, representing much of the trading element of Timbuktu. The Moor, she wrote, was:

> One of the few people in the world who, however much his life and circumstances may vary, never seems to alter in the least respect. Here, many miles from his own country, he is the same shaggy, wild-eyed star gazer, the same bit of biblical silhouette.

Apart from the needs of trading, the Moors did not mix with other races but sat in the marketplace 'in little groups, silent and self-centred. Never do they look prosperous or well-fed.' Their women were aloof, 'drawing their blue garments across their faces against the glance of a stranger'.

Dolly was interested in the Sonhrais women, who had varying degrees of freedom depending on their class. The wives of important men had far less liberty than lower-class women and lived 'in an almost cloisteral seclusion'. The wife and sister-in-law of a local judge told Dolly that except to visit relations they had not left their house for many years. When she asked if they got bored, and in answer to their questions told them of the busy lives of European women, they stared at her blankly.

Dolly had found that among many women in primitive cultures, 'the great luxury of a "lady", and [one] that every good husband tries to give his wife, is the luxury of inertia. To think, to work, to take exercise, is the unpleasant necessity of the poor!' So far, she had only once encountered a 'feminist' in West Africa. Near Mopti, she met a woman of the Peuhl (or Fula) people, who worked in service for a family. Her married life had begun with beatings from her husband who was also mean with money. Taking matters into her own hands and defying his wishes, she went into service and made much more money than he. Now he dared not beat her nor deceive her, she told Dolly, because if he did, she would leave him and take their three children with her, and then he would have to work, which he did not want.

The 'supreme feminists', she discovered, were the women of the nomadic Tuaregs. In *The Road to Timbuktu*, Dolly devoted a chapter to this ancient and extensive tribe, members of a sect of Islam about which her knowledge appears impressive, although she says her illness limited the time she spent with them, which made her study 'superficial'. Theirs was (and remains) traditionally a hierarchical society, with nobles, vassals and slaves, although Dolly pointed out that 'slavery in this case has little of the unpleasant significance it has for us', for the captives were well treated, largely because they were objects of value and a source of wealth. Dolly was impressed by the women's position, which would 'promote green envy in the heart of the most pronounced suffragette, the most "modern" woman in England. She has a voice in all public and family conclaves and she is completely the mistress of the tent and her children.'

The Tuareg woman also had the power to divorce her husband, and if she was discovered to be unfaithful, the most a husband could do was remonstrate with her and not retaliate on her lover. However, although they had the benefit of their men being monogamous, among upper-class women plumpness was prized to make them attractive for marriage and Dolly wrote of daughters of nobles from the age of 7 being 'subjected to a fattening process almost cruel in its severity', being force-fed with milk and starchy food and not permitted to take exercise so that eventually they could not move without the assistance of slaves.

She was invited to a Tuareg encampment by a chief who was 'friendly to French civilisation' (many were not), and vividly described the circle of men sitting immobile on their haunches outside the tents like birds of prey in their dark blue, almost black robes and veiled up to the bridge of the nose, which gave them an air of aloofness and mystery. The women had many questions for Dolly about her married life (as would many she met), if her husband beat her and whether there was divorce in England. They thought it scandalous that a man could divorce his wife – and for mere infidelity – and yet were surprised that after seven years of marriage she herself was not yet divorced. They did not believe Arthur loved her because she was so thin – if he did, he would instruct his servants to fatten her up.

She left the camp musing on the religious and superstitious beliefs of the Tuaregs and the ever-present *djins*, spirits that inhabit the Earth and take the form of humans or animals. As she did so, a tornado blew up, the hot wind shrieking 'like a tortured soul' and blowing sand in her face. Even her guide was frightened, 'not of anything obvious but of the *djins* that he was convinced were fluttering around us'. She considered the people she had just left: 'Truly children of the storm are the Touaregs, born of the violence of the sun and the sand [...] masters of desolation.'

Dolly learnt of many superstitions among tribes while in Timbuktu. Some believed in the power of charms and sorcerers, who could foretell the future, bring rain, exorcise an evil spirit or even will a person to death using an astral body. Others feared death and the existence of *djins*. For some, bad luck lay in having a house opening to the east, for others in seeing another tribe's blood, which meant they may not fight each other. When she 'cured' an elderly man's headache, not by magic but with aspirin, he was so grateful he gave her an amulet into which he had sewn a powerful *gris-gris*, or charm, which he assured her would keep evil spirits from hampering her travels. He warned her not to open it or the spell would be lost, and she obeyed him, even though she suspected the *gris-gris* to be simply 'a bit of giraffe skin, or possibly a leopard's claw or a twig off a sacred tree'. Nevertheless, that she had kept it closed indicated the power of suggestion that such talismans could exert over the wearer.

But while these aspects were part of what she termed the 'medievalism' of Timbuktu, she was also aware of steps being taken to bring it into the twentieth century, one of which she happened to see in action and was an occasion when she was in the right place at the right time. The post of administrator, held by a Frenchman, had always been a military role but while she was there, the first civil administrator was installed. To witness such a momentous occasion, all the chiefs of the neighbouring tribes came in from the desert, riding into the city decorated with flags, while locals and Europeans alike found vantage points to watch. Dolly thought it an impressive pageant, with officers on horseback heading the cavalcade containing the new administrator (Monsieur Léonce Jacquier), accompanied by the commandant 'and a cohort of Touareg chiefs on camels, with floating cloaks, lances and bucklers, and a host of other natives on small, spirited horses'. Tom-toms, singing and drums concluded the short but memorable event.

Dolly also saw history in the making with the arrival of the Citroën Mission, an experiment by the French carmaker André Citroën to make crossing the Sahara possible in a fraction of the time it took a camel. The new *chenilles* were caterpillar-like vehicles with tank-like wheels, and Dolly had seen two of them in southern Algeria the previous year as they prepared to make their momentous trial trip. Eventually, they had rolled into the finishing point at Timbuktu on 7 January 1923, thus completing the first trans-Saharan journey in a motorised vehicle and changing the face of desert travel.

Although she had arrived too late to see that first contingent, three more of the original models arrived during her stay, to be greeted with much noisy excitement by the world's press. While she appreciated the progress they represented, she thought they stood out in the natural environment as 'stark and inharmonious, almost obscene', and they could do little to change the existing limitations of desert travel – the lack of water and petrol supplies and facilities for reprovisioning. It made her think again of René Caillié, whose year-long journey a century earlier to the gates of Timbuktu 'runs like an epic of endurance and heroism',[15] although there were no journalists to herald his arrival, no money spent on him or encouragement given. Now the heroes were 'heroes of mechanism – squat, ugly, malodorous, invulnerable, grunting through sand so thick that nothing wheeled, not even a hand-cart, plies in Timbuktu'.[16]

She was surprised that the local people took little interest in the 'caterpillars'. That vehicles should propel themselves without visible powers did not surprise them, as they saw it as just another example of the ways of white men, 'whose every action and appurtenance is incomprehensible, and without apparent reason'. It was aeroplanes that really alarmed them, she said, for anything that could pass swiftly through the air like nothing human must surely have 'some unholy connection, some diabolical agency'. At that time, great progress was being made in coping with the various difficulties of aerial transport across the Sahara caused by the intense heat, and Dolly was assured that it was the aeroplane that was 'going to supersede the great cruel roads of the centuries, to wrest their birthright from the sons of Ishmael [and] to violate that great, arid virgin, the Sahara'.

Yet she was aware of the risks. Throughout her books, Dolly highlights figures who were little known to the world but deserved recognition, one of whom was François-Henri Laperrine. A French general during the First World War, working among desert tribes in the Sahara, which he had crossed many times, he pioneered a flight in 1920 from Algiers to Dakar. Dolly had sat with other tourists at Biskra, anxiously scanning the sky for signs of his aeroplane, which they never saw. He had to make a forced landing and survived seventeen days with a broken leg, before dying of hunger, thirst and fever in the desert within just a few kilometres of a well he knew, for in that desolate area there were no landmarks.

It was time for her to leave Timbuktu, for the weather would soon be turning unbearably hot. She had been in West Africa for three months, two weeks of which were spent in Timbuktu. Having discovered 'the chief natural pests of West Africa', which were the sun and mosquitoes, she was learning how to cope with both. Quinine was the great saviour for those suffering or trying to avoid malaria and she was disappointed by the 'anti-quininites who are spreading a rapid and feverish propaganda' and who got 'as heated over it as the Irish question [such examples probably being the modern equivalent of anti-vaxxers and Brexit] or Lloyd George'.

Despite her periods of illness, which included dysentery, she had accomplished a lot, although less than she would have liked. She wanted to stay longer but there was no overland route: the last convoy of the year had left, and the water level of the upper Niger was getting low. Her only choice was a *chaland*, a boat the size of a small fishing vessel, flat bottomed and made of iron, in which she had about 10ft to herself to eat, sleep and live in. The rest of the boat was occupied by ten *laptots*, African colonial troops who were a vital part of the French administration and from whom Dolly screened herself off by means of a thin grass mat. A contraption she made from sticks, bamboo poles, a bit of canvas and string formed her storage area and ceiling, but even though she was petite, there was little room to stand up, so she kept bumping her head.

To add to her discomfort, she had to wear a sun helmet from dawn to dusk, even when inside, for the sun stole in everywhere, the temperature reaching 130 degrees in the shade.

Yet gradually life on the *chaland* settled itself into a sort of routine, improving from 'unadulterated Hades' to 'squalid discomfort' to 'almost comfortable, almost homelike'. Still, the journey posed constant challenges. The crew went on strike when they believed a man she had taken with them was a sorcerer because he was of another religion. The fastest they could travel was 2.5mph and they often struck rocks or got stuck on sandbanks and were plagued by flies during the day and mosquitoes at night. Every ten minutes, she had to place wet towels over the sides of the boat to avoid being scalded. From midday to 6 p.m. daily, 'one sat literally in a bath of perspiration', which made her excessively weak, and even at sundown when a little breeze sprang up, she still had 'violent attacks of perspiration' which made her feel she had no blood left and made her tense at the slightest movement. Saghair was no longer with her, so she bought her own food, eggs and chicken when they passed through villages and did her own cooking.

Later, on the hottest day of all, they crawled into the village of Koulicoro, where she was given the luxury of 'a sit-down luncheon in the bungalow of a kind Frenchman in a world that no longer rocked or grunted or squealed'. Next day, a train took her the short distance to Bamako, where she found the population of 400 white inhabitants and '"civilisation" [...] monstrous, almost frightening'.

She had to wait there for five days for another train but was happy to take 'bath after bath at the house of an angelically kind young Englishman', the head of a British trading firm. She wanted to return to the coast through French Guinea, but already, in March, it was too late in the year. After three days' journey from Bamako by train and car, she arrived at Dakar, and three days after that she was at sea, 'with a salt-laden Atlantic gale blowing fresh health and strength into every tired limb, and nerve and muscle'.

Looking back on Timbuktu, she summarised it succinctly: 'Murder, rape and pillage, gaiety, culture and commerce, slaves and perfume, all these made up the personality of the faded beauty who lies brooding, half-sleeping under the intent eyes of her last and faithful lover, the sun.'

On leaving, she had felt 'a queer regret' at turning west, and as she hammered the nails into the sticks on the *chaland*, her heart ached. It seemed to her that she had not reached the road's end but only a junction:

> … where the road forks east and south; [I felt] that I was turning my back on it […] that still it ran on triumphantly, clearly, tempting me, beckoning, through the fever-laden rivers of the south, through dark forests where even the fierce sun of the Equator can scarcely penetrate […] where the air hangs heavy as a miasma, into the heart of the great Dark Continent, to the nethermost fringe of the Never-Never Land. I felt that farther – much farther maybe, or perhaps only round the next bend of it – something still waited for me.

She had to find out what it was.

MONSTER WOMAN

Dolly arrived back in London on 14 April 1923, her return anticipated with interest by all sections of the press. Many hailed her as the first British woman to reach Timbuktu, others as the first English woman to do so, a few even as the first white woman,[1] but whichever was the more accurate, the importance to her would be 'first' and 'woman'. When interviewed by *The Times*, Dolly said the experience was worth the ordeal, but she did not recommend that anyone who was afraid of 'sordid discomfort' attempt the journey, except at the one brief favourable time of year.[2]

At home in Ebury Street, she was reunited with Arthur, who was back from Argentina with several published articles to his name about ranching. A photograph was published of them together, both holding rifles while she showed him a leopard skin she had brought back.[3] Their reunion was an occasion to celebrate not only each other, but more achievements, because during the previous year, amid preparation for their travels, they had each completed another work of fiction, both published by Duckworth and out on the same day. Arthur's *The Primrose Path* was a series of fourteen short stories with 'cosmopolitan' themes. One review of his talent said he had 'a lively and vigorous pen, a knowledge of the east and an understanding of the ways of the world'.[4]

Dolly's novel *The Road* was clearly influenced by her earlier travels in Tunisia and by the 'Sheikh complex', as she dubbed it. Lisbeth, a young woman who lives in a dull town on the south coast of England with her abusive husband, is strongly attracted by the call of 'the south' and Africa, and senses she may have roots there. After divorcing her husband, she marries an older man, an Arab, and they live in Tunis, but he soon dies.

A French diplomat falls in love with her but is murdered by a tribe. Her ex-husband finds her, stalks her and is in turn murdered. Lisbeth is captured by the head of an Arab tribe who has loved her from afar and has discovered that she is part of his tribe, who was stolen as a baby. She falls in love with him too, and allows him to declare her his bride, even though she has realised that he is the murderer.

One review said the psychology was not convincing, but Lisbeth was real, whether one liked her or not, and the desert scenes were well done. *The Bookman* said, 'Our eyes are blinded by the sun and by its sandstorms […] we are fascinated, hypnotised, by the horror and mystery of it.'[5] Dolly would surely be pleased by the review that said she had 'succeeded in striking a note of originality which makes her book stand out among novels of the kind. There is a vividness and a vitality about it, which are remarkable.'[6]

She was now busy with commissions for features on her Timbuktu expedition, as well as writing *The Road to Timbuktu* and in between, she slipped back into the familiar social scene. Her bons mots continued to be quoted, her attitude to the relationship between the sexes having been strengthened by her experience of managing without a partner while away: 'A man, as man, is no longer essential to woman […] Marriage, materially speaking, obviously handicaps a woman's career more than it does a man's.'[7]

However, tough words did not mean she did not love Arthur. In July, they left for the Continent for six weeks, spending time in Picardy, where they intended to do some writing. It was a pattern that would endure: travel separately, then reunite and socialise; holiday together, preferably somewhere 'quiet and almost dull',[8] so they could write; return and socialise; write.

Her creative output was extraordinary. Another of her short stories was published in the 1923 Christmas edition of *Magpie* magazine, alongside Agatha Christie. In December, she and Arthur frolicked at the increasingly fabulous Three Arts Ball, for which she was once again an organiser and posed for the press. This time, she was veiled up to the eyes in a striking Egyptian costume, which was rather apt for the year that had seen the opening of King Tutankhamun's sarcophagus. But although she enjoyed getting back into stylish clothes, downing cocktails and smoking cigarettes, she needed her contrasting worlds:

> It is good to feel that one has left one's little niche in the everyday world, where each one of us is assessed and tabulated to a nicety, to slough off one's everyday accepted self, and to lose oneself in the anonymity of a strange country and people, among whom one has to make good solely by the leverage of one's personality and will to win.[9]

Dolly left for another 'strange country' in early January 1924, this time with her father, with whom she must have reached some sort of truce, and his wife Gladys. She was joining the couple on a voyage to the West Indies, disembarking in Jamaica, from where she intended to make the short journey to Haiti, about 100 miles away. She was going on her own, as usual. Before she left, she commented that, as a rule, women did not make very good travellers when it came to the 'uncomfortable and arduous' business of getting around, and unless she could go with such intrepid companions as Rosita Forbes or Charlotte Cameron (an American explorer), she preferred to go solo.[10] Once again, she and Arthur coordinated their departures and the day before she left, he sailed for New York to go on to Brazil in search of material for his next adventure book.

Still little known to the outside world, Haiti was of interest to Dolly because it was the world's first black government republic, having secured its freedom from centuries of French colonial rule and slavery in 1804. However, it had become increasingly unstable, which was a source of concern for the US government which, because of Haiti's location, had long expressed a diplomatic and strategic interest in its stability.

After seven of its presidents were assassinated or overthrown between 1911 and 1915, the USA intervened in several ways. The resulting Haitian–American Treaty saw the creation of the Haitian Gendarmerie, essentially a military force made up of US citizens and Haitians and controlled by the US Marines. The USA gained complete control over Haitian finances, and the right to intervene in Haiti whenever the US government deemed necessary. However, some of the Gendarmerie's policies caused unrest, as did a new pro-American president, who was foisted on the country, leading to a revolt in 1919–20.

After investigating claims of abuse, the US Senate reorganised and centralised power in Haiti. By the time Dolly arrived, it had achieved some stability, and a select group had achieved prosperity, though most Haitians remained in poverty. She wanted to learn more about what she called 'the civilising aspect of black humanity' and afterwards, she would admit that her experience caused her to 'shed a good many preconceived prejudices'.[11] To the outside world, Haiti tended to be seen as Dolly herself had once described it, 'a comic-opera republic',[12] manned by officials with overblown titles wearing pompously elaborate uniforms. 'Praise of Haiti was an unfashionable thing,' she wrote. 'No one wanted to hear the truth, papers would not print it and people would not read it.'

Her destination was Port-au-Prince, the capital, and for such a relatively short distance from Jamaica, getting there was unexpectedly difficult. There was no regular boat service to the island and the only way of reaching it was by Dutch cargo boats, whose arrival times were random, so it was a question of waiting and hoping that an anticipated boat did not decide at the last minute to miss out Haiti altogether and head for Cuba.

She had to wait around in Kingston for nearly three weeks and when she did get on a boat, she was the only passenger other than a Dutchman in the margarine trade, who proved (usefully) to be not only a polyglot but courteous and amusing company. 'We became firm friends, and he fussed over me like a motherly old hen,' she wrote. 'I think he regarded me as some kind of amusement clock-work toy.'[13]

When they reached Port-au-Prince one evening six days later, her Dutch friend helped her deal with the noisy and incomprehensible bureaucracy and eventually she managed to find accommodation of the standard officially reserved for white visitors, predominantly Americans, in a wooden verandahed villa. While she was relieved that she was staying in an annexe rather than the main house, where bedrooms were barely separated by flimsy partitions, she found the washing facilities challenging, consisting of shallow pools in the garden, each with 3ft of water 'that might or might not be changed every day, in which everyone bathed in turn' and in which small greens frogs chased each other along the bottom.

Although such conditions were rather better than those she had endured in the desert or African bush, one advantage of living primitively in those places was that of anonymity. However, she soon found to her horror that it was not to be the case for her in Haiti. Before she left Britain, she had spoken to the press about the reasons for her visit.

Over half of Haitians were Roman Catholics but still adhered to Voodoo (which had originated in Haiti) from their slave heritage, when they had been forbidden to practise it. The roots of Voodoo lay in African religions in which were incorporated elements of Christian symbolism. Dolly wanted to study this aspect of their belief system, but it was said to be difficult to penetrate.

Although that was not the only reason for her journey, a few modest lines in the British papers were seized upon by the American press, where they 'burst forth and screamed in flaring headlines'. Now those papers were laid out before her in the office of every important official she went to see whose cooperation she had hoped to secure. 'EARL'S DAUGHTER GOES TO SEE BABIES EATEN IN BUSH,' screamed one. 'BRITISH SOCIETY LADY SEES HIDEOUS HAITIAN ORGIES,' shouted another.[14] 'NOW DAINTY LADY DOROTHY SEEKS VOODOO'S DREAD SECRETS,' headlined *The Meriden Daily Journal*.[15] And all of them were accompanied by unlikely photos and followed by paragraphs, even pages of amplification. While there was no excuse for such sensationalism, there was no denying the impact that Voodoo, or at least the perception of it, had on the collective imagination, with tales of black magic, orgies and human sacrifices.

Offended by what they saw as her damaging image of the island, the Haitian press itself had not been idle either, copying examples into its own papers and adding comments, labelling her 'A monster woman, with a revolver in one hand, a vitriolic stylo in the other'. She was called an 'ogresse' and a *'mangeuse de nègres'* (negro eater), which made her cower 'under a tornado of sarcasm and hate'.[16]

The mischief of the American press had prompted a fishing expedition, whereby enquiries were made of Dolly, reference books and old newspaper articles dredged up and a complete dossier compiled on her, 'past, present and imaginary'. When she protested that it was all an exaggeration, she was asked if she was prepared to talk to the Haitian press and spent an evening and much of the next day answering questions from a selection of journalists in which she was 'cross-examined and re-cross examined' and made to talk in a combination of French and

English. 'Traps were laid for me and bait cunningly trailed, to see if I would not give myself away, if I would not make rash admissions of any of the things laid to my charge.'

She found they did not easily believe her sincerity or open-mindedness, and she did not blame them, for they said journalists had visited the island before and, behind a mask of friendliness, had traduced and ridiculed them. As a journalist herself, Dolly counter-charged them with believing what other journalists wrote, and for writing against an unfortunate woman without checking if what they said was true: the appellation of 'monster woman' had really got to her. Eventually, she made headway, and one journalist apologised for what now seemed the absurdity of the insult, and they even managed to joke with her about her revolver.

Eventually, they were friends, the men's hostility transformed into an interest in her purpose and a willingness to help her understand their island and their struggle to present it 'with fairness and justice to a sceptical world'. These Haitian journalists impressed her with their education and intellect, as well as their stylish dressing. She felt there was no malice in the Haitian, for although he was quick to curse, he was equally quick to bless – in her case, to eat his words. Now headlines appeared that were complimentary about her. The morning after their meeting one paper, *Le Temps*, headlined, '*la voila l'ogresse, elle est charmante!*' (Here's the ogress [female ogre], she's charming!), while *La Post* called her '*Une gentille exploratrice*' (A kind explorer).[17]

In Port-au-Prince she found 'broad streets jostling with parti-coloured humanity, with men of every shade, from coal black to pale olive, dressed in tropical European suits, and dusky peasants from the countryside in loose cotton rags'.[18] There were French, Dutch, German and Spanish, too, who had married into Haitian families, as well as American marines and some of Chinese ethnicity, who ran cafés and small stores.

Dolly found much to like and to admire about the city and its people. She found that the Haitian ideals of politics, society, art and literature were French. The formal and precise speaking of French by the upper and middle classes impressed her, for it was clearer than that often heard in France, while Creole, spoken by the lower classes, was 'more a patois than a language' and mingled several European languages with some

African. She saw something of the arts in Haiti, represented by poetry, sculpture and music, 'all three the natural arts of the African'. During *Mi-Carême*, carnival week in Port-au-Prince, she discovered just how passionate the Haitians were about music and dance, and she learned the *Merengue*, 'a kind of syncopated tango, irresistibly lilting'.

However, not all was sweetness and light. While she was there, six editors were in the prison for being too critical of the government. The recent President of Haiti, Monsieur Louis Borno, whom Dolly met, was unpopular because many Haitians considered him a puppet in the hands of the Americans under their high commissioner, of whose rule he was allegedly supportive. She was taken to the prison to see the men but was not allowed to talk to them. She did establish that they were treated slightly better than other inmates, although some had been waiting for months for an indefinitely postponed trial.

She also learnt that 'although the Haitian is proud of being racially black he has a curious colour snobbery of his own'. Upper-class Haitians tended to be lighter skinned than those lower down the social scale; a light-skinned girl would not willingly marry a man much darker than her, and vice versa. This lightness of colour, said Dolly, came from former times, when the French 'mingled much with their slaves'. Some girls enhanced their paleness by discreetly applying face powder.

Overall, however, Dolly's impression was very positive. 'I saw none of the comic opera I had been led to expect of Haiti,' she wrote. 'I saw a little world that was just like any other little tropical colonial world one knows, except that it was black.'[19] Nevertheless, she recognised the challenges that the island faced, 'And yet under all the laughter, one sensed that there ran a current of discontent, of distress, the distress of a people who knows itself at a disadvantage, who feels that its liberty is in peril'. The people had a 'covert hate of its white overlords', because the island with its untapped riches was coveted by the Americans, 'who, in their turn are a festering sore to a people who would rather make a mess of their country in their own way, and retain the freedom of their souls'.

She wrote poignantly of her sympathy for Haiti and the pride of its people, particularly the 'stress and struggle that made them into the first negro republic in the world, their country into the cradle of a black hope'. In the voices of young militants, she heard the anger and frustration of helplessness, and she wrote of the 'tragedy' that underlay the land, one that had been 'turned into a farce by the white races, because of a matter of skin pigmentation'.

Those she met entreated her to write about them fairly; they did not mind constructive criticism, if she also gave them credit where it was due. They wanted to prove they were as good as white people. They wanted her to write about their faults, their virtues and their struggle to rise in the world.

Part of their story inevitably included writing about that other matter that had got her into trouble – Voodoo. However, despite all the contacts she had made, and the warm hospitality shown to her by Port-au-Prince society, she found it extremely difficult to persuade anyone to talk to her about it, for which she blamed the American press and their silly stories about her.

A form of nature worship which involved many gods, historically it had sometimes taken the form of animal and sometimes human sacrifice too, particularly a child – 'the Goat without Horns'. The priest was the *papaloi*, rather like the African witch doctor, who traditionally wielded his power through his knowledge of drugs and 'medicines'. The use of charms played a big part in Voodoo, while dancing was its outward manifestation, so much so that when Dolly arrived she found the US authorities in Haiti had commandeered all the drums they could find that were used in the ritual.

There were some people who were willing to talk to her but only in secret. Dolly was invited to a house in a remote part of the island arranged by a friend, but it was mysteriously vetoed 'in high places'. An anonymous letter that was slipped under her door, which she showed confidentially to an official, was never returned to her. At least one letter threatened her if she continued to make enquiries.

Then late one evening, when she was out dining with friends, she got a message to meet someone immediately at his house, for which he sent a car. He had heard of a ceremony that was to be conducted that night in the hills by one of the last influential *papaloi* in Haiti and he had arranged for Dolly to watch it unobserved. For two hours, they drove uphill on an 'execrable' road, at the end of which a guide waited with horses for them to ride 7 miles across loose, stony ground, which took another hour and a half. Dolly was feeling extremely uncomfortable, as she was still wearing a flimsy evening frock, having had no time or opportunity to change.

Eventually, they reached a clearing with a low wooden shelter in front of it, in which she and her companion were able to conceal themselves and peer through a gap. Twenty yards away was a large mud and plaster hut with its double door open opposite them. On one side of the clearing a fire shed a flickering light 'on some fifty or so negroes, dressed in loose white garments, who revolved in a circle with rhythmic tread, stamping their feet to the low throbbing of a tom-tom'. There were six or seven women, also in white, a couple of whom were young but the rest Dolly thought were probably over 50, and all the men were middle-aged; she thought that by day they were probably labourers and artisans. Their expressions were calm 'but set and tense, wearing an air of expectancy, of pent-up excitement'.

They were waiting for the *papaloi*, who was also wearing white with a red cap made from a knotted handkerchief on his head and no adornment except a dull-coloured bangle. A small, wizened man, Dolly put his age at around 65. As he entered, the tom-toms burst into loud drumming and then someone handed him a fluttering white cockerel, which he whirled around his head, then 'wrung its neck, letting the blood drip into a small bowl on the ground. Its body he flung to the crowd, who hurled themselves upon it.' He then entered the mud hut, from where Dolly could see him pouring the blood into other bowls and adding something to them, which she thought was a sort of powder.

When he emerged, bowls of liquid were handed around from which the participants drank greedily. Dolly thought it must be an alcoholic drink, 'judging by the energy with which they literally hurled themselves into a wild dance. Round and round they went in wide circles, stamping and shaking,' while their leader kept up a monotonous chant and the music got louder. 'They kept it up interminably, untiringly, never varying the movement or the rhythm.'

After forty-five minutes, a girl fell to the floor as though exhausted but she got up when someone stumbled into her and continued dancing. Dolly thought the dancing was more rhythmic and regular than ordinary 'roadside' dancing and appeared to have a concerted plan. Sometimes, a dancer whirled away from the circle and others followed, passing so near to her and her companion that they could feel the breeze of their garments on their faces. There was lots of wild spinning, and as more drink was passed round, the chanting and stamping gained vigour.

A man raised his voice in a long howl that sounded like a wolf, the sound echoing eerily down the silent hillside. Sweat poured down the faces and arms of the dancers:

> Flecks of foam stood at the corners of their lips, and their fever-bright eyeballs protruded. Once they paused as a girl with a shrill cry started spinning like a top. With a sweep of her arm she tore off her cotton blouse and cast it aside. Five minutes later her skirt followed suit, and round and round she spun, naked in the moonlight, looking like a wild black goddess of Paganism.

At the end of an hour, another chicken was killed and later a small pig, whose throat the *papaloi* cut with a long knife, his acolytes holding the creature while its blood gushed into a bowl. In their excitement, 'some of the blood was spilt, making shiny scarlet pools on the black ground, smearing their bodies and their white robes,' she wrote. 'Each time a sacrifice was made, the blood was carried into the sanctuary and every time the *papaloi* re-emerged, the rejoicings broke out more fiercely and unrestrainedly.'

Dolly found the monotony and the repetition mesmerising. 'The muffled throbbing of the tom-tom sent the blood beating fiercely through my veins, making me feel as if I too must leap up, and scream and dance.' Everything started to feel unreal: she found it incredible that they were only about 100 miles 'from the world of tourists and the great trans-Atlantic liners and I dreamed I was back in the heart of the great Dark Continent six thousand miles away'.

At around 5 a.m. the ceremony started to wind down, the tom-toms becoming sporadic and the dancers tiring. Some fell and stayed on the ground. The drinking bowls were empty and there were no more sacrifices. 'It all looked like a complicated clockwork toy that is on the point of running down.'

By the time Dolly and her companion got back to Port-au-Prince four hours later, she was shivering with cold and fatigue, with her dress torn and a shoe missing its heel, but it was worth it. She knew it was 'white' Voodoo she had seen and she was unable to say whether isolated cases of 'black' Voodoo ever still occurred, although she seems almost to hope that they did, as evidence that the 'dark soul of Africa' continued to survive despite the encroachment of civilisation.

The contrast between Dolly's refined upbringing and petite stature and the physical and psychological demands of her expeditions would always fascinate, and while she was often written about with admiration and even awe, the fact that she was attractive and female was, in the eyes of many (mostly men), her vulnerability. Despite remarking that her exploits so far 'would have daunted most men adventurers', a feature in *The Meriden Daily Journal* ended patronisingly, 'Veteran American marines are wondering if their next job will be the rescue of a pretty English noblewoman from the hands of voodoo witch doctors'.

In fact, it is likely that she was the only white woman, and probably one of very few white people of either sex, ever to see a Voodoo ceremony. Thirteen years earlier, a journalist for *The Times*, reporting from Haiti on the revolution, said he (openly) witnessed a Voodoo ceremony (in which there were 'no cannibalistic features' but women put their hands in the fire and beat their heads against a tree), and noted that 'it is seldom that a white man gets the chance of attending these ceremonies'.[20] Always willing to reassess her views, she found that 'If travelling teaches one nothing else, it teaches one to chuck all preconceived notions and prejudices into the sea, and to believe nothing until one has seen it for oneself'.[21]

After a month in Haiti, she returned to England in April and had to deal with the press. Most papers wanted to know only about her Voodoo findings. She made it clear that she had seen only a few animal sacrifices and that the stories that circulated were grossly exaggerated, with the result that at least one paper concluded that she was 'disillusioned' by it. Her feature for *The Sketch*[22] emphasised the burgeoning culture and the Haitians' fervent hope for the future, 'when they will stand on an equality with the white'. She hoped her accounts would contribute to a wider understanding of the island.

Only her mention of their resentment of the 'intrusion' of the USA hinted at tension. Her feature was illustrated by a range of photographs she had taken: the president and his (white) wife with other officials;

the local all-black football team; the bustling market square at Port-au-Prince with its glorious cathedral, and a peasant's hut in the interior; and as a reminder of the island's political situation, a group of American congressmen being entertained by Haitian society. In July, she spoke about her experiences on the new medium of radio on a station called 2LO, better known as the BBC.

Her views on Haiti were expressed more forcefully in a feature in *The Washington Times* in May 1924, apparently from an interview with her in London. While she acknowledged the Americans had done much in the way of road building and sanitation, Haitians had told her they would like to have more American schools and libraries and would welcome American enterprise to help them exploit the country's resources. The country was very fertile, and yet 'it grows hardly enough to feed its own population. Except for a few of the more fortunate class the people are poor.'

Nothing had been done to aid the republic's industrial development and given the huge sum Haiti owed to the USA, she said, it would be in that country's interest to encourage industrial prosperity, otherwise Haiti 'will not be able to pay [it] in a century under present conditions'. The American marines themselves were not popular and 'the bulk of the natives detest them and tolerate them only through fear'. In addition, many Haitians considered the military force kept there by the USA to be larger than necessary. 'They think that Uncle Sam is not so much desirous of maintaining peace in Haiti as he is of maintaining a military foothold in the island that might be strategically valuable in the case of trouble with Japan or any other country.'

In her absence, reviews of *The Road to Timbuktu* had appeared. Widely and highly acclaimed, the book would cement her reputation not only as a fearless female explorer but as a travel writer. This 'deeply interesting' book much impressed the reviewer of *The Scotsman*, who was moved by 'varied sympathies and thrills of admiration in following the writer through the perils, the illnesses, the mishaps and the victories of her brave and lonely journey [… She has] an alert eye […] and a pen that can justify the childlike wonder that reads romance into the ruthlessness of the jungle.'[23]

For now, though, it was time to reunite with Arthur, whose travels in Brazil had furnished him with fresh adventures for his stories and interesting souvenirs, particularly the rattle from a rattlesnake, which he said was an effective charm against other serpents. Like Dolly, he was enjoy-ing reviews from his latest novel, *The Yellow Dragon*, based on his travels in China. 'A thoroughly entertaining yarn of adventure and intrigue in Hong Kong and China,' said the *China Express and Telegraph* on 15 May. Although it considered there were too many characters who should be subsidiary to the central figures (a French Vicomte and a young woman), it pronounced it amusing and enjoyable.

Later that year, another of Arthur's short stories, *The Grey 'Oss*, would be published in *The Bystander*. Of the success of him and his wife, the *Daily Mirror* said that 'both are reaching individually the enviable "best seller" class. Lady Dorothy has the Walpole talent in her blood, and Mr Mills is a writer of directness and charm.'[24]

It was not often that Arthur's comments appeared in the newspapers – usually they were Dolly's – but when *The Yellow Dragon* came out, he told a reporter in their Ebury Street flat that if anyone doubted he had been to China, they just had to look at the furniture in one of the rooms, almost all of which reflected his travels. In the decade of new dance crazes, and having returned from Brazil, he also told a joke about the Brazilian tango, or *maxixe*, in which an elderly father angrily throws out his daughter's young man when he finds them entwined in each other because, deaf to the gramophone, he does not realise they are dancing.

Arthur's own preferences do not seem to have extended to that activity, not even with his wife, although Dolly was happy to do her own thing as always. After returning from Haiti, she was spotted at the Carlton with a younger man, Captain Ulick Brown, a glamorous and talented fashion designer, where they demonstrated a dance that Dolly had recently discovered.

That Arthur and Dolly had no children clearly freed them from the attendant obligations and left them with time and opportunities they would not otherwise have had, yet even so, what they managed to fit into their lives is impressive. A novel Dolly had somehow managed to write during the previous year was published. *The Arms of the Sun* was an adventure-thriller about world domination, set in Africa, with strong science-fiction elements such as a genetically engineered man-monkey (who defies his villainous creator and dies saving the heroine and her suitor) and other monstrous creatures, as well as futuristic biological weapons (phials of plague). 'So thrilling that it takes the breath away,' said *The Daily Mail*,[25] although one review said that while it was exciting, it was reminiscent of H.G. Wells' science-fiction novel of 1896, *The Island of Dr Moreau*.

She had competition, too. In the *Illustrated London News*,[26] a review of her novel appeared above (surely better than below) that of the latest book by the American author of the *Tarzan* novels, Edgar Rice Burroughs. His *Pellucidar* was also a science-fiction adventure which took the hero from the Sahara into the Earth's crust, where he finds a world of strange creatures. Nevertheless, Dolly's book, with its female heroine (an explorer), arguably had wider appeal to women, who might not usually read the genre.

As if she had nothing else to do, in July she played the part of Queen Anne in a three-act play called *Queen Anne's Orangery*, written by Mrs Ernle May and staged at St George's Hall in Bloomsbury, in aid of her mother's favourite charity, Our Dumb Friends League. She had not ceased writing about society, noting for one feature the emergence since the war of both 'social pirates' and 'the New Rich' and how the latter provided lucrative opportunities for the former – apparently charming but unscrupulous men (for it was always men) – to make money, in part by preying on unsuspecting middle-aged women.[27]

The Mills' adventurous year culminated as it had done previously in costumed splendour at the Three Arts Ball, this time at the Royal Opera House. Dolly's co-organisers included the President of the Arts Club, Queen Victoria's granddaughter, Princess Marie Louise. Hundreds of partygoers, augmented by actors when the theatres closed, waltzed and fox-trotted around a splendid fountain whose spray danced and splashed to the music, yet without getting anyone wet because the 'water' was made from 10,000 tiny balls made of a phosphorescent material. An electric rainbow, sunshine and silvery moonlight all added ambience, through which a procession of 'royalty' passed, wearing gowns heavily studded with 'jewels' made from pieces of red, white and blue glass and silver and gold paper. Dolly was a judge of the best costume of the night but soon would swap her own for more practical travelling garb, for the drummer was calling again.

A MIDDLE EASTERN MELTING POT

Almost two years to the day after Dolly began her long and challenging journey to Timbuktu, Thomas Cook's travel agency began selling tickets to the ancient city, thanks to the Citroën Trans-African Company and their revolutionary desert vehicles that she had encountered. One paper said the itinerary sounded like a mad dream: there would be a service twice a week, taking the traveller from Paris to Timbuktu in twelve days, across the Sahara and French Sudan, with big-game hunting available along the banks of the Niger. The first service was due to leave on 4 January 1925 and it was rumoured that royal personages would be among the passengers, with Madame Citroën, wife of the millionaire promoter, joining them.

To her shame, Dolly became possessive of the places she had explored that were becoming public property: 'Some of the fine flavour is lost, some of the fervour of the cult.'[1] She recalled being at a luncheon party where she was introduced to a 'large and opulent lady' who had just returned from Timbuktu with one of the new tours, enjoying luxury hotels and liveried chauffeurs. '"Oh rotten place," she murmured languidly. "The car broke down for several hours, and it was so sandy. I wish I'd done it the easy way you did."' Until Dolly 'remembered to laugh', she chewed her lunch 'with mild rage' as she recalled her own tortuous journey before the existence of roads and six-wheelers.

But there were still many places to explore that commercialism had yet to find. As that first load of tourists went to Timbuktu in their special vehicles to discover that sandstorms left nothing untouched, Dolly left for the Middle East. Arthur, as ever coordinating his itinerary, left for Brazil.

With a cheap ticket to Constantinople[2] in her pocket, she left Victoria Station on a drizzly January morning in 1925, anticipating the realisation of a dream 'of treading the Golden Road to Samarkand [... of] losing myself among the great Tartar Hordes, whatever they may be, way up near the Chinese frontier, of making a pious pilgrimage through the country of the Arabian nights and of reaching the Caucasus'. She also fancied buying an Astrakhan coat 'at the place of its birth', but that Russian city, on the banks of the Volga, would remain out of reach. The necessary permits from Moscow were not forthcoming, much to her regret, and the Middle East generally was not conducive to exploration. In the end, she left without any kind of itinerary or much hope, just 'a sneaking trust in the great bearded god of nomads who has generally befriended me, ill and tired and depressed'.[3]

Her itinerary would be ambitious and potentially dangerous: Asia Minor, Syria, the Holy Land, Transjordan and into Iraq. Changes to borders and regimes in the Middle East since the Great War had created further hostilities and uncertainty. Even travelling through Europe was not edifying. The jolting train seemed to stop everywhere, and then it either rained or snowed 'and all the stations bore a cousinly resemblance. Ostend, Brussels, Cologne, Nuremberg, Vienna loomed and receded', punctuated by incessant visits from ticket and baggage inspectors and so many countries to pass through that she never seemed to have the right language or money.

Catastrophe fell after Budapest, when it was realised she was on the wrong train and destined for a place for which she had neither ticket nor passport. The train staff 'gathered in mass formation and, with the pomp and volume of sound of the Heavenly Host ejecting Lucifer from Celestial regions, cast me out into the snow near the Roumanian frontier'.

She spent two days lost, boarding 'all sorts of ridiculous trains', and one night slept on a nest of suitcases she had constructed in a corridor. Constantinople, when eventually she reached it, was cold and wet and she was tired and ill and nothing could cheer her: the cobbled streets and myriad steep steps hurt her feet; she had chosen a bad hotel; the city was not as she had imagined it; all the smart officers wore galoshes and no one wore a beard. Everything was just wrong.

Dolly knew her weaknesses: when she was cold she was 'a pitiable object, abject and mummified, without initiative or courage'. More than anything, she was 'sick for the sun', the quickest way to which was due south via Asia Minor (also called Anatolia), 'a route that everyone tried so strongly to dissuade me from taking that I promptly took it'.

Her journey was made at an interesting time of great change, and she would be entering what was, in effect, a new country. Built on the ashes of the ancient Ottoman Empire, which 'slumped across three continents',[4] the Republic of Turkey had formally come into being less than two years earlier, on 20 October 1923. The revolution that gave rise to it came after the Great War, in which Turkey had supported the losing side.

Here was a new secular republic 'in which the sultan's subjects would become modern citizens, the age-old privileges and discriminations based on religion and sex would be abolished and wealth would belong to the people'.[5] Under President Mustafa Kemal Atatürk, reforms in clothing, which had long been the distinguishing mark of rank, gender and religion in Ottoman society, sought to forge equality between men and women and also to blur the differences between Europeans and Turks. For men, one new freedom lay in no longer having to wear the fez, and for women, not being compelled to wear the veil (most Muslim women would abandon the heavy face veil but choose to keep their head covered).

On the way to Angora,[6] Dolly started to feel happier. Her pleasantly heated train ran along the shore of the Sea of Marmora, through fertile lands of vegetables, apple orchards and olive groves, the sky turning the Gulf of Ismid a deep blue and the Islands of the Princes that studded it resembling those of a fairy tale in the shimmering sunlight. In the distance, the summit of Mount Olympus was powdered with snow.

Angora itself was cold, as she had expected in winter, but the sun shone and there was no wind. Dolly delighted in the intricacies of its ancient mosques and, as always, in meeting the women. She found the new freedoms that were coming into being had not yet been adopted by the women of Angora, unlike their sisters in Constantinople. Despite their new president encouraging them to appear in public life freely, they kept themselves apart, spending their time in each other's houses, and were seldom seen on the streets and never in a café or cinema. The manager of Dolly's hotel told her he had tried to get tea dances going, in the hope that his customers would bring along their wives, but instead, they came on their own, which rather missed the point.

Those women she spoke to, who she described as lower and middle class, did not seem too bothered by their segregation. They reminded her of other women she had met elsewhere, who could not understand her love of travel and her desire to write and earn money, asking her why, in that case, had she married? Surely, it was for the husband to make the money, so that his wife could enjoy the finer things in life. Dolly's attempts to explain the joys of being gainfully occupied and having economic independence were invariably met with incomprehension.

The woman in the provinces, she noted, was not well dressed, wearing a shapeless black garment, half-hood, half-cloak that was 'neither Oriental nor European in its conception'. By contrast, the higher-class woman of Constantinople was often dressed in the latest Parisian fashion, encouraged by the reforms. Dolly felt – 'though I say it with timidity' – that, on aesthetic grounds at least, she had 'made a mistake in discarding the undoubted allurement of the veil', for her beauty often lay in her fine eyes, whose expressiveness was 'cunningly enhanced' by kohl.

When she discussed the subject of the unveiling of Turkish women with an Arab friend in the northern Sahara, he said his view was that it was not good for a virtuous woman to show her face, and although his wife told Dolly that she too would keep the veil, it was (she said, confidentially) for a very different reason: 'If we no longer wore the veil, how would we deceive our husbands?'[7] Being unveiled meant everyone in their small town would know where she went and how long she stayed, but as things stood, she could move about largely unrecognised and if in doubt, she could borrow a friend's garment for further disguise. This gave Dolly, who was aware of the symbolic oppressiveness of the veil, much satisfaction whenever she saw 'the self-satisfied flourish with which the Arab carries abroad the enormous key that he has just turned upon some unprotesting, secretly-smiling wife!'

While most women still played a silent role in Turkish life, she paid tribute in her resulting book, *Beyond the Bosphorus*, to one who had played a vital part in her country's political rebirth – Halide Edib Hanoum, a novelist and activist, 'whose voice cries with perhaps a more powerful note than that of any woman in our emancipated north'. Dolly's own observations of women's lack of interest in their low social status was an issue of which Hanoum was very aware.

However, some reforms were eagerly adopted. When Dolly expressed surprise at not seeing any men with beards as she had expected, she was gravely told, 'Democrats do not wear beards' and noted that 'nine out of ten Turks are as clean shaven as Americans'.

The ideal place to people-watch was at the Fresco Hotel, the nucleus of social life in Angora where, under the new regime, Cabinet ministers were happy to mix with coachmen and where she bumped into a couple of British acquaintances she had not seen for years. Nevertheless, she was one of only two women who entered Fresco's during her stay.

In fact, Dolly said she was 'one of the first European women to visit Angora' and she must have been one of even fewer women to write about it. She would have been aware of the older Grace Mary Ellison, a British journalist who spent time in Turkey before the Great War and did much work for women's rights. She had returned in 1922 to cover the Turkish War of Independence and wrote *An Englishwoman in Angora*, which took a partisan look at the political situation.

By contrast, as an explorer and general observer, Dolly tried to avoid bringing politics into her writing. Nevertheless, she did allow herself some comments on the new regime. She thought Turkey was 'trying to run [before it had] learned to walk alone'. She had hoped to meet the new president (a comparatively young man in his early forties), but he was in another region. She considered him 'undoubtedly a very brilliant man [… yet] though his power is nominally but that of President, his power is that of Dictator'.

Being one of few women made her an object of keen curiosity. The Turk, said Dolly, was 'one of the most courteous men in the world' and his curiosity about her 'was rarely obtrusively expressed, and nowhere was I stared out of countenance, or annoyed by small impertinences, as I have so often been in other countries'. She had long since learned that saying she was a writer helped to explain her presence, and saying she was a journalist was even better, seeming to guarantee her respectability.

In Angora, she took a public bath where, after undressing communally (with women who had 'the very worst figures I have ever seen in my life'), she tried to wash alone but was commandeered by a large, muscular lady who cleaned her roughly until she looked like 'a curried prawn', then she was beckoned to join a group who were roasting nuts and eating sweetmeats and wanted to know all about her, the first European woman many had seen at close quarters.

Turkish food gave her different levels of enjoyment, such as 'cheeskobab' – 'small lumps of meat swimming in sour cream which, after the first shock, was not as bad as it sounds' – and the sweet halva, which she loved. She was thwarted in her search for Angora cats, famed for their fine and silky coats and unusual eye colouring, and all she found were emaciated examples.

It was at some point during this journey, apparently while she was still in Turkey, that she stopped at a place whose name she omits for reasons pertaining to the safety of others and had a curious and rather haunting encounter. She needed to write a piece for which she had been commissioned and stayed for a while in a 'semi-Kurdish village' high in the mountains, above the ruins of ancient Turkish fortresses, where she was invited to take her meals with a local Kurdish family.

Borrowing an old mule called Mustapha, she explored the valleys and mountainsides and one evening noticed a tiny pinprick of light high up. Her friends avoided answering her questions as to what it was, saying it was probably a shepherd's light. In daylight, she rode up to the place, stopped in what seemed to be an old garden and found herself looking down the barrel of a gun. On the other end of it was 'one of the most strange and striking figures I have seen in my life'.

It was a young European man, who instantly made Dolly think of a picture of the Archangel Gabriel she had seen in a gallery. Dressed in ragged brown breeches, his blue shirt widely open at the neck, he was tall and slender, 'with a pale fine featured face [...] brushed-back golden-red hair, whose close tight curls receded from his temples, and his eyes, rather close together, were of a curious greenish blue'.

Once she had apologised for trespassing and he realised she presented no threat, he said he was a recluse. He invited her into his log cabin and as they started to talk, albeit awkwardly, she became aware that he was 'a highly-bred gentleman and a man of the world', who spoke of

London, Paris and New York, and while he would not reveal anything about himself, she decided he was Russian, an impression he confirmed when he said everyone called him Ivan. At that time, there were many Russians from good families who had been driven from their homes after the revolution and murders of the imperial family.

When Dolly asked her Kurdish friends more about him, they continued to be evasive and said it was not good for her to be Ivan's friend. She got the impression that he was a Russian of high birth who had greatly suffered in his country and harboured bitter hatred of the new regime. Two years earlier, he had built his hideaway and took great steps to preserve his privacy.

Naturally, her friends' warnings did not stop her and she took to having tea with him in his cabin in the late afternoon. Still he gave nothing away of himself and lacked 'gallantry or sentiment', maintaining a curious detachment. Even when he remarked on the blueness of her eyes (which many did), comparing them to gentians in the field, it was without any note of admiration or compliment. However, he had clearly lived a cosmopolitan life, was cultured and artistic and 'one of the most interesting conversationalists' she had ever known. When it was time for her to continue her journey, his detachment continued: they parted unemotionally without any suggestion of a future meeting.

Only on her homeward journey did she find out more when she stopped to see her Kurdish friends again and asked about Ivan. They said he had left a month earlier, taken away by someone but they did not know who. Unknown men had called on him in the middle of the night and after that he had disappeared; local people found his cabin burnt and the garden trampled to nothingness. A rumour reached them that he was dead. They suspected he was involved in a political conspiracy with his own countrymen.

Two days later, Dolly was in another town dealing with a bureaucratic matter when she was approached by a southern European who had helped her earlier. He warned her to be careful who she spoke to about 'Prince X' (Dolly does not reveal his name), and when she said she did not know him, the man said she knew him as Ivan.

In a hotel in Constantinople, where she was arranging the last part of her journey home, the head waiter gave her a letter which he said had been delivered by a young man in poor clothing who spoke in a French accent and pointed Dolly out. The letter was sealed with a signet ring engraved with a coronet or crown, which she had seen before. Dated two weeks previously, it read, 'When you receive this, Madame, I shall be dead and I am glad. Thank you for your blue eyes.' She learnt nothing more, recalling it as 'the queer but half-conjectured tragedy that I stumbled upon in that melting-pot of intrigue and violence that is the Middle East'.

From Angora, where she had stayed a week, she continued by train through southern Anatolia towards Syria. After being locked in the waiting room of a busy station by an over-cautious stationmaster who took the safety of lone females very seriously, she was much troubled by bureaucracy, her passport seemingly of great interest to every new military policeman who boarded the train. Not having the language, she sometimes resorted to asking passengers in other compartments if they spoke French or German and getting them to act as a translation service until the policeman was satisfied.

The railway ran east through the beautiful Taurus Mountains, then sharply south to Aleppo. The builders were contracted to take the line as far as Baghdad in Iraq, but in Dolly's time, engineering challenges meant it ran only to Nissibin, on the Turkish–Syrian border, about 100 miles short of Mosul and several hundred miles short of Baghdad.

As the mountains grew less rugged, they approached the plains of Syria, where she noticed that every little wayside station had its name in French now as well as Turkish, a sign of very recent events. Syria had been part of the Ottoman Empire; in September 1923, as part of its partitioning, France was assigned the League of Nations mandate for Syria. Dolly captured the colourful blend of humanity that peopled the stations:

[A] small sprinkling of dapper French officers, and sleek Levantines and hooded Arabs mingle with the extraordinarily picturesque Turks and Kurds of the mountains, dressed in baggy trousers and bright handkerchiefs, sturdy robust creatures with rugged bronzed faces.

As always, she turned her observations about this journey into compelling narrative. Harsh facts are wrapped in lyrical prose and descriptions of brutal events are softened by the relating of associated myths and legends. She sees into the soul of a place and its people and finds romance in the histories and struggles of those who have called it home or who strive to do so, and yet she is never sentimental.

Humour is never far away. On her first day in the much-anticipated city of Aleppo (which, to her disappointment, remained totally devoid of sun), she made only one diary entry, 'Why did I send my clothes to the wash? It will keep me here until Tuesday.' She spent an enforced three days there, observing the complex mixture of races and tongues and cultural tensions set against a riot of colour and movement and noise. As always, she notes the women. While the eastern Mediterranean woman was more emancipated than her Turkish sister, she was 'more rigidly kept than in Europe'. The men were 'jealous and mistrustful to the last degree' and whatever modern or feminist ideals they may have got from European men, they found them hard to apply.

In the Holy Land, her first stop was Jerusalem, whose essence she distilled:

At first sight, Jerusalem charms the eye and the imagination; on closer acquaintance, one only sees her infinite sadness. She is sad because her beauty lies in tradition and in sacred associations and now all her beauty, all the glory of her heritage, is shattered by the vulgarity of man.

The only biblical place that seemed not to be profaned was the Via Dolorosa. It was not surprising that 'Jesus wept for Jerusalem, wept for its wickedness, its blindness, its heartlessness. Today one could weep over its vulgarity, its hatred, malice and uncharitableness. This is from its religious aspect. Historically of course, it is interesting'.

She talked to different types of Jews, demonstrating her knowledge of their history, and to pilgrims, as the many tourists liked to call themselves, whom she viewed cynically and blamed for taking up all the decent accommodation. Only when she went on to northern Palestine (with a government official as her knowledgeable guide) and at Tiberias stood by the shores of the Sea of Galilee did she feel she was truly on sacred ground. Crossing the River Jordan, she was overwhelmed by the beauty and perfume of the wildflowers, as they drove past Jewish shepherds, nomadic tribes and young *chalutzim*, or pioneer farmers, often from the ghettos of Europe and America, who sought to build new settlements.

On she went to Nazareth and later Nablous, where she and her guide encountered a 'relic of a dying race', the tiny colony of Samaritans, living in their ancient sanctuary that dated to 700 BC. Regarded as heretics and frequently exiled, the Samaritans had never married Jews and there were few women among them. In 1925, there were thought to be only 200 Samaritans left in the world.

Dolly liked Acre, 'just a little town out of an old fairy tale, a fortress tiny but fierce', which was visited by St Paul and had withstood the battering rams of the Greeks and Arabs, 'of Saladin […] Richard Coeur de Lion and Napoleon'. Acre was 'a romance' for her, until she visited the prison, where, after spending an hour, she felt as though 'all the sunshine of the day had been blotted out'. The worst and most violent criminals of Palestine were sent there, where the existence of the death penalty was evidenced by a set of gallows and where she found two prisoners, 'poor lost souls', waiting on death row.

Her spirit was restored after a drive up Mount Carmel to the stronghold of the Baha'i movement, whose followers Dolly found to be 'a lovable and fascinating people […] idealists who have dreamed a dream of peace that passes all understanding'. Near Haifa, she was reminded of the recent tragedy of a young Jewish woman, Sarah Aronsohn, who witnessed part of the Armenian genocide by the Turks, which had made her join the Nili ring of Jewish spies working for the British in the Great War. When she was caught and brutally tortured, to avoid the risk of revealing any names, in October 1917 she shot herself, but lingered on in agony for days until a doctor friend agreed to help her die. She was 27.

Having headed south, she stopped briefly at Jaffa, which was like 'a city of the dead' apart from the presence of Palestinian police and army officers, who were preparing for a significant event taking place not far away at Tel Aviv. It was there that Dolly again witnessed history in the making, part of a story which remains relevant in the twenty-first century, sometimes tragically so. She was to see the arrival of Britain's Foreign Secretary Lord Balfour, who would open the first Hebrew University on 1 April 1925.

Her powerful description of his visit must rank as one of the most engrossing accounts of the time. Britain had conquered Palestine from the Ottoman Empire during the Great War, after which British rule was administered under a League of Nations mandate. In 1917, the Balfour Declaration gave land in Palestine to the Jewish people as a national homeland alongside the Arabs. Now, tens of thousands of people crowded into Tel Aviv to see this man. As Balfour's car approached, they went wild, shouting and crying, faces transfixed:

I have attended a good many emotional public functions in Peace and in War among men of all kinds, creeds and colours, but never [...] have I seen, heard and actually felt such a fierce fervour, such a concentrated force of mass joy and enthusiasm.

She felt it was no exaggeration to say, 'I think so must our ancestors have acclaimed their Christ.'

It was dangerous at times. The sheer density of humanity saw her gasping for breath and nearly trampled underfoot, while the hostility of Palestinian Arabs which had threatened to disrupt Balfour's visit was ever present and expressed in the headline of their leading newspaper, la Palestine: 'The Death Knell of all Arab Hopes'. Anti-British feeling ran high.

She was given an official tour of Tel Aviv which, although she admired its industry and growth, she considered ugly, like a 'mixture of Margate and Port Said, with a dash of Hollywood thrown in', and she was permitted to attend official functions during the formalities and celebrations of Balfour's visit. Jewish children sang, religious leaders gathered, the Catholic Archbishop of Damascus was spotted near a group of inscrutable Arab sheiks, and speeches were given in English, Arabic and Hebrew.

Yet for all the joy and hope, Dolly was concerned for the future. The Arabs and Jews, although both Semites, were 'of such widely differing temperaments and mental planes' that they would always 'hate each other, as only blood relations can do!' She concluded, prophetically, that the new situation could be creating 'yet another knot in an already sufficiently tangled world [...] Only time and in particular the next few years will prove.'

The contrast with the next part of her expedition was marked. 'An abomination of desolation' was how she described the Dead Sea Valley, 'where death has laid his cold, withering hand, a reminder that nothing is fixed, that our planet will one day be but a dead ice-cold star'. Smugglers favoured it as a place to shelter with their contraband, with the lack of respect for the law continuing in Jericho, where there was nothing to do except rest and where an Arab friend of Dolly's, an official for the British Government, beat a hasty retreat when she and another bureaucrat were due to dine with him, fearing his knowledge of 'midnight ambushes and shadowy flittings' would get him into trouble.

Body and soul were refreshed when she took a strenuous climb to visit 'a little colony of Greek monks marooned in mid-air' in an ancient monastery, a warm and hospitable group of men who invited her and her guide to sit on hard, knobbly chairs 'that looked as though they had been bought in Tottenham Court Road', in a room with gaudy carpets and walls hung, incongruously, with pictures of George V, Queen Mary and the Queen of Holland.

The journey from the Dead Sea Valley to Transjordan (now know as Jordan) was not one to be undertaken lightly 'if you are squeamish about hairpin turns, greasy narrow roads and dizzy heights', where a wrong move could see a vehicle being hurled over high precipices into rocky valleys, 'with not enough left of you to make a decent funeral'. After two unhappy hours up the Mountains of Gilead, they reached Es Salt, still remembered as the place where in April and May 1918, Australian and British troops had bravely fought the Turks.

The Jordanian people she found handsome, picturesque and dignified. Even in the wild hinterland, Christianity had found its way. Dolly was not a fan of missionaries, and she was reminded of the reason when she reached Madaba, a colony of Greek Catholics with a mission school run by Syrio-Arab teachers. 'Without cynicism or disrespect, I wonder why missionaries of whatever race or denomination always succeed in making the God they teach so unattractive.' In a school room made to look as dull as possible, beautiful young women:

> … minced sedately in clothes of horrible European influence, taught to
> sing Methodistical hymns in the nasal whining that is the Arab method

of song, in English of which they did not understand, and with their fine strong hands built to milk a cow [...] or ease the weariness of a tired brown warrior, taught to make odious little anti-macassars and doyleys of drawn threadwork.

Amman, the capital, was the first stop on the railway that ran south to the holy city of Medina, where Dolly saw pilgrims of various races passing through. Many had been forced to take the 'cruel roads', walking or riding in lorries, on camel or mule, often with no word of the language of the country through which they passed, 'buoyed up only by their infinite faith and stoically meeting death sometimes by hunger or thirst, by cold or by treachery'.

She encountered the Wahabis, 'that predatory, recklessly brave tribe of nomads', whose ruler Ibn Saud had recently been in the British press in a situation that exemplified what Dolly called 'a tangled network of Arab politics'. In the hills above Amman's railway station stood an important base of the Royal Air Force, a lonely posting, because the men were not allowed to take wives or children with them to this unsettled area. Nearby was the home of the British-trained Arab Legion, a mobile fighting force comprising around 15,000 men, who patrolled the desert and kept an eye on the Wahabis.

She continued travelling towards Iraq, where she did not have permission to enter but decided she would face the consequences once she got there. She wanted to investigate the Yezidi tribe, whose reputation as devil worshippers made them an oddity and contributed to their persecution. They were to be found in an area north of Mosul in northern Iraq. The best way of getting there to avoid official red tape, as well as many miles of stony desert, was to go from Aleppo in northern Syria, still in French territory. There she arranged for a ramshackle Ford car – 'all the best people cross deserts in Ford cars nowadays, except under exceptional circumstances' – and a Christian Arab chauffeur, Jacoub. As she did not have much luggage, the French car owner asked if she would mind an Arab family occupying the back seat for part of the way, a motley group consisting largely of women and one child, who kept falling on top of her.

Interesting encounters during stops for water, repairs and meals included those with officers of the Foreign Legion, some of whose members she had met before and who she found dispelled the myth that they were recruited entirely from criminals or outcasts. While it was true that many were violent, she had empathy, even admiration for them: 'They are men, not brutes, and in all my varied experiences in Africa [...] I have found many a worse gentleman than the Légionnaire.'

As they drove on, Jacoub pointed out that the rust-coloured streak that cut through a sand dune ahead of them was the blood of a chauffeur who was shot by bandits the previous week and told Dolly, rather too brightly for her liking, that he hoped they would not hear that an English person was travelling through and carrying money, for they always liked to stop the English. By now, she had discovered that the 'family' travelling with them was an Egyptian theatre troupe who were on their way to performances in Deir ez-Zor and later Mosul and who distracted her entertainingly with their stories and idiosyncrasies.

Dolly was not impressed by the behaviour of the Arabs in this area, agreeing with the view expressed by earlier writers that, unlike the Bedouin and other nomadic tribes, the effect of the sedentary life of cities and villages and restrictions of farming communities had been a deleterious one which brought out 'all that is vicious and ignoble'. She experienced this when staying in a local hotel and, unable to lock her door, propped her baggage against it. In the middle of the night, a scuffling sound woke her and in the moonlight she saw a man climbing through her window with a knife in his belt. Hurriedly, she aimed her revolver at him and he disappeared. Without sufficient luggage to barricade both door and window, she slept rather lightly for the rest of the night.

It was not the only unpleasantness on that journey. Half a day was lost complying with a bureaucratic issue when she had to wait for the necessary official to get out of bed. Then, when trying to get the car over the Euphrates River on a wooden raft, a crowd of thirty or forty Bedouin jumped on, which overcrowded the raft and saw many people jumping overboard. Some swam but others were swept away by the current and one man drowned not far from Dolly. The Ford broke away from its moorings and she feared she would be crushed by its lurching. Eventually, the boatmen managed to steer the raft into shallow waters and the remaining passengers waded ashore.

As Dolly dried off on the riverbank, she saw someone make off with one of her bags which contained her paperwork, and she had to threaten to shoot him if he did not drop it. They managed to get the car going again but then, in a day that seemed doomed to misfortune they hit a sandstorm, which was the worst she had ever known:

> There was nothing on the vast horizon to break its violence, a burning *khamsin* [gale] blew that scorched our faces, making the skin feel as if it were being flayed with a rawhide whip, cramming the sand down our throats and ears and eyes, stifling us, filling every crevice of our clothing and making all but our tinned food inedible.

All sense of direction was lost, yet it was safer to continue than stop, with Jacoub driving inch by inch. It was a nightmare which she hoped never to experience again.

The next day, with their water levels very low, they continued towards Badr, which roughly marked the border with Iraq, but were warned by a band of nomads that in the hills ahead lurked bandits who the night before had robbed a traveller. Dolly knew this was the riskiest part of the route but there was no other way they could cross the hills.

A short distance beyond, they spotted several heads watching them and a gun aimed at them. Calling to Jacoub to put his foot down, she crouched on the floor of the car as he bumped forward as fast as he could at the same time as the brigands started making their way towards a point at which they could block them off. As the car came level with them, two shots rang out but fell far short. Minutes later, they were out of the hills and into the open desert, safe for now.

FROM LEOPARD MEN TO
DEVIL WORSHIPPERS

Discomfort and hunger continued even when they reached Badir because there was no accommodation to be had, and Dolly was very tired and felt ill. As they left the next day, she was slightly cheered by the pair of tall, muscular and splendidly dressed patrol guards who accompanied them for protection until they reached the official frontier of Iraq, where Dolly met up with her little theatre troupe after a performance en route.

They were not far from Mosul, which she had been eagerly anticipating but, during a brief break at the interesting but permanently troubled Tel Afer, she felt sick and dizzy and realised she had sunstroke. The troupe were to travel in convoy with her, but when they made excuses that they could not leave until that evening, when Dolly wanted to get away earlier, she found herself, most unusually, in tears of frustration and weakness. Five aspirin and half a flask of brandy later, she somehow reasserted herself and they were off. However, her first sighting of Mosul three hours later was lost to her forever in a confusion of 'heat and pain and faint sickness'.[1]

Here, Dolly's evasion of official formalities caught up with her, and she now had to deal with the loathed bureaucracy and spent only a day and night in Mosul. Nevertheless, she knew its history and that of its ancient sister, Ninevah, one of the greatest cities of the ancient world, whose remains stood on the opposite bank of the Tigris. For millennia Mosul had been a strategic location due to its crossroads and bridge between north and south, east and west, making it attractive to different cultures. Among those who had besieged it, wrote Dolly, were Saladin, the Mongols and the Persians. Less than a century after her visit, it would become known to the world for its devastating occupation by ISIL/Daesh.

Even while she was there, she was aware of 'a sub-current of strain and tension. For Mosul is a problem of perplexity. We do not want it, but the Turks do, very badly.' Guerilla warfare between the Turks and the Kurds was a constant. In 1924, a dispute had arisen whereby Mosul was claimed by Turkey, a situation opposed by Iraq alongside Britain and its mandate. The League of Nations ruled that it should be returned to Iraq, which Turkey accepted. However, Dolly feared the situation might spark 'more than a political war, a religious war between east and west, between the great Moslem world and Christianity, and cost the universe a deal of life and money'. As she drily put it, 'Northern Iraq is by no means a health resort or rest cure'. It was 'largely unpoliced and teemed with wars and rumours of wars, tiny wars, not worth space in the papers', but small events might fan into bigger conflicts. The situation required 'the most careful nursing'.

On leaving Mosul, she had planned to visit Jebel Sinjar, one of the chief strongholds of the Yezidi, but an incident there involving tribes 'that never in the history of the British occupation had been known to disagree' delayed her departure until the following day. 'Aeroplanes carrying bombs were rushed out, officials [...] dashed out in the middle of their dinner and to my regret my car was kindly but firmly commandeered.' It made her muse on how it was more by luck than judgement that she had not got herself into real trouble roaming the desert without proper protection.

It soon transpired that the fighting between the Turks and Kurds in the Jebel Sinjar area was worse than usual and the authorities at Mosul had 'expressed themselves very clearly' on the matter of her journey. Instead, she had to content herself with a study of the Yezidi plain dwellers in the village of Bashika, their principal stronghold outside the war zone.

Historically, much mystery accompanied the complex religion of the Yezidi, who rubbed along peacefully with Christians (Catholics lived among them), who they regarded as fellow sufferers for religion's sake. However, it was a different story with their Muslim neighbours, who hated and persecuted them and, tragically, their persecution continues in the twenty-first century. Once a very powerful race, the Yezidi were thought not to be 'masters of a book', seemingly having no written doctrine.[2] From being wild avengers of the wrongs carried out against them, over time they had become more subdued and less mistrustful, although Dolly still found herself restricted in what she could see and find out.

She was fortunate to be invited to stay in the main meeting house in Bashika, where visitors were rare. There she was graciously received by the leading man of the village, a Kurd, and by the elderly and charming Syrian Catholic priest. She was then taken to meet the local chief of the Yezidi at his house, which had a large courtyard with a welcome stream of ice-cold mountain water running through it. In the shadow of large, plastered pillars, she and her English interpreter joined the chief and a dozen or so of the leading Yezidi men, who took it in turns to sit in the place of honour opposite the chief. Many of them puffed on large, curved pipes, the chief offering his to the guests, although Dolly preferred 'a chain of my own cigarettes' as she watched them drink innumerable tiny cups of tea.

She noticed how good-looking all the men were, 'with bronzed open faces and honest eyes', and the build of mountain dwellers, even though they were plainsmen. They were mostly dressed in white with red turbans, except the chief, whose headwear was black, and their manners 'were dignified and quiet, but they looked as if they could fight: the sort of men one would like to have in a tight corner!' Their original language was a Kurdish dialect, although the plains people had come to speak a type of Arabic. Her interpreter told her they were mostly discussing local affairs and the state of the crops. Their food, she discovered, was similar to the Arabic, but unlike Muslims, the Yezidi drank wine when they could get it.

Her brief stay in Bashika was spent viewing the sacred buildings with their curious conical, fluted spires and talking to whoever she could, but although everyone was very charming and happy to answer her questions, frustratingly, the one thing they would not talk about was 'the forbidden subject of their religion'. While the Catholics were willing to tell her everything they knew about it, their knowledge was not very detailed, and she had to look to other sources.

Although the Yezidi were referred to as 'devil worshippers', she ascertained the term was misleading in their complex belief system, which was a compound of many others, mostly Christian, with a dash of Judaism and Islam. They revered God (Yazdan) and believed in the archangels (with Satan at their head) and in the Second Coming of Christ (Melek Isa). However, Yazdan was too kind and merciful to be effective, whereas Shaitan (the Devil) was quick to anger and be vengeful, so had to be appeased. When the world was created, he was cast from Heaven to Earth over an argument about the creation of the snake and founded the Yezidi religion. His reign on earth lasts 10,000 years, of which there were 4,000 left to run. At the end of that time, he will be restored to the hierarchy of heaven, when the reign of evil will end and Melek Isa will rule for another 10,000 years.

Dolly learned that at their yearly festival they offered up one sheep to Melek Isa but seven to the Devil. His symbol is a peacock-like bird, a metal image of which was kept in their main sanctuary at Sheikh Adi, which no man of any other faith had ever set eyes upon, while a jealously guarded clay replica was kept at their temples. His name must never be mentioned and when he is referred to, it is by the equivalent of 'Emperor Peacock' or 'the Mighty Angel'.

Dolly discovered there were many taboos within their beliefs, including the colour blue, the mention of fish or lettuces and the mispronouncing of certain sacred words, the breaking of which could lead to death. She was fortunate to be shown around the Yezidi temple by the High Priest himself, 'a really beautiful old man with a long beard, dressed in snowy-white flowing garments and turban'. Through arches at the end of a courtyard she could see the shrine, empty except for a low altar, at which she gazed 'with suppressed longing and curiosity' for she knew that behind it was kept the clay replica of the Sacred Peacock, 'surely the most curious satanic emblem in the world'. She was ashamed to realise that she would have betrayed her hosts' trust and taken any risks to catch a glimpse of it, but the doors were locked day and night, and she was 'no cat burglar!'

However, her stay coincided with the Yezidi New Year on 16 April, and she would later tell the British press, whose interest was particularly piqued by the 'devil worshippers', a story that does not appear in

Beyond the Bosphorus. Once a year, at their greatest feast (presumably that of 16 April), the clay figure of the peacock was taken from its secure place and shown to worshippers inside the temple. Dolly said that, with the help of her chauffeur, she managed to climb up and peer through a window to see 'a crudely carved, greenish image'[3] before which everyone kneeled. Perhaps she decided to omit this episode from the more permanent record of her book for fear of criticism or reprisals. For the New Year, the temple and the saints' shrines were decorated with flowers, including – unusually – one dedicated to a female saint, Ziti the Beloved.

Dolly managed to visit a few ordinary Yezidi women. They did not veil like their Arab neighbours, but were treated more considerately by their menfolk who, although they were allowed to take four wives, rarely did so for economic reasons. Not for the first time, she found the women totally without curiosity except in small personal or domestic matters: the Yezidi woman was an 'untutored daughter of Nature'.

When she left Bashika, wishing she could stay longer, she was delighted to receive many kind messages and compliments and was touched that half the population seemed to gather to see her off. They themselves were flattered that someone 'from beyond the desert and beyond the seas' should visit them. She found the Yezidi 'simple and kindly folk, Devil Worshippers, yet broadly speaking, in many ways so much more Christian than the Christians!'

On her homeward journey via Mosul, Dolly stopped at several little villages of Syrian Catholics dotted about over the great plains and in her book talked of the history of Christianity in the region, including the fabled king, Prester John. He had ruled his kingdom in Asia around the twelfth century and was said to have fought for Christianity against the Muslims and 'idolaters' over much of the Earth's surface. She walked among ancient crumbling churches where shepherd boys carelessly played, and the ghosts of long-dead believers whispered their stories of worshipping behind armed walls.

Back in Anatolia, she found a war zone, 'a sheer nightmare this time, of dirt, discomfort, suspicion and semi-starvation'. A place she particularly disliked was Adana, which was notorious for the massacre of 1909 whose victims were mostly Armenian and who were also deported from there in 1915. She felt the town's hostility to foreigners

– she was briefly but painfully stoned – and said it was not a place whose memory she cherished. Later, the European cities flashed by once more, but this time they looked good to her, 'for there was health and joy in my heart'.

Dolly's energy and drive was further evidenced shortly before her return when the novel she had completed the previous year was published. *The Dark Gods* was, as the publishers, Duckworth, said, 'the story of a lost soul in Africa', which she also adapted into a play. *The Scotsman* described it as 'an interesting story of outpost adventures in a very remote and very unsettled village in West Africa', to which a Frenchman takes his young wife, but she falls in love with a native man and becomes a victim of negative forces.[4]

Arthur came home before Dolly, with his latest novel *The Gold Cat* also published. Set on an Argentine cattle ranch, his fast-paced adventure story with a love interest was 'A rattling good story', said *The Birmingham Post*. 'Mr Mills evidently knows the country he has chosen for his setting.'[5] He was also writing features on Argentina for *The Bystander*.

Dolly spoke publicly about their separate travels, something she was often asked about, which made her wonder if it was of more interest to editors than her adventures themselves. They were, she said, 'a semi-detached couple'. She found it amusing when people expressed their astonishment about their travelling separately for three months every year. 'We are both writers, of different temperaments, different style, different moods and different themes,' she explained. They wandered off on their own ways, avoiding the worst part of the English winter and returning in time for the start of the season.[6] Even on holiday together, they did some work, but their working patterns were different, so they often did not see each other until the evenings. Arthur was an early riser, but she was not, and her imagination was better after lunch. When asked if she got lonely without him, her response was:

Not a bit of it. After all, we are only away from each other for three months of the year and we can always make up for the absence when we get back home again. I always travel alone. My only companions are occasional guides or porters.

If Arthur or any other companion was with her she would get no work done, she said, and her time would be wasted, and Arthur apparently thought the same. 'How would I soak myself in local colour and customs?' Dolly asked. 'The atmosphere would be lost.' Inevitably, they would always end up discussing 'the everyday problems which husbands and wives discuss when they are together'. She thought it an excellent idea for spouses to holiday away from each other occasionally. Above all, 'Alone, I am able to mix with the peoples of the countries I visit: live their life in their own surroundings and learn their customs'.

Whatever Dolly's preferences about doing things on her own, the success she and Arthur were enjoying from their writing saw them being invited to many events, jointly and separately. In October, they attended a literary event at the American Woman's Club in London hosted by its president, Caroline Louise Brown, wife of the founder of the famed literary agency Curtis Brown. In her writing, Dolly mentioned having an agent and it is possible they were from here, hence her invitation. Among the clients who attended was D.H. Lawrence, recently back from Mexico; he was recovering from a serious attack of malaria and could have shared with Dolly horror stories of tropical diseases.

At around the time she returned from the Middle East, Rosita Forbes came back from an expedition to Abyssinia (now known as Ethiopia), which she declared would be her last. Recognition of the achievements of these 'great women explorers', as the press referred to them, was increasing, but could be ruined by condescension – too often, they were seen as bored society ladies looking for thrills. A feature in an American paper pitched Dolly and Rosita against each other in a 'who will win the lady explorer's catch-a-sheikh-or-a-cannibal contest?'[7] – an indignity that their male counterparts were spared. When the two were spotted at Ascot, there was musing as to how Mrs Forbes, looking pretty in a soft-brown lace dress, could possibly have tackled Abyssinia.

Sometimes, it took a female journalist to point out how they differed from men. The reason women were willing to exchange home and friends for bandits and mosquitoes, said one journalist, was the urge for change which leads them to embark on adventure. Unlike men, they needed no definite objective. If a man takes to the bush, he is 'prospecting for diamonds or iron, or going to build a railway.

Perhaps he has gone out to hunt big game or butterflies, or preach the Gospel.'[8] If adventure follows, he enjoys and sometimes pursues it, but for the most part, 'men consider the proper place for adventure is within the pages of a novel'. The only exception, she said, was Sir Ernest Shackleton, but otherwise women were 'much more venturesome than men'. While Dolly's choice of destination might be prompted by a particular sphere of interest, above all, she simply liked to take to the road.

As she saw out the year as usual at the Three Arts Ball, this time dressed as a bat in black veils – vampires were a popular theme in the 1920s – she already knew where she was going next: Liberia. She knew it would not be easy, and it was probably the expedition for which she needed the most assistance, both with officialdom and practical matters.

Liberia was an independent republic, the only country in West Africa which was not a colony of any European power. It had been settled in the nineteenth century by the USA, mostly with freed US slaves from Barbados and America, whose descendants now lived along the coast and held the reins of authority. After being neutral at the start of the First World War, Liberia had joined the Allies and came to international notice. However, it was still a difficult country to travel in, because the heavily forested terrain had not yet been surveyed and the maps that did exist seemed to have been compiled by hearsay.

On 2 January 1926, Dolly sailed from Liverpool to Monrovia, the capital city. Unusually, she did not publicise her destination in advance, saying merely that she was going to West Africa and would return in April.

Permits were not easily obtainable and letters of introduction were necessary, for which she had a letter from the Liberian minister in London to the Secretary of State in Monrovia. She also had a letter of introduction to the President of Liberia himself, Charles D.B. King, whom she wished to meet.

As she approached Monrovia, nestling on a green hillside, Dolly thought it 'must have seemed a land of promise to those first weary settlers bowed down but not broken by generations of slavery'.[9] On arrival on 19 January, she presented herself and her letters to the British Chargé d'Affaires at the British Legation, at whose house she stayed while formalities were dealt with and who gave a reception in her honour. He informed the Secretary of State of her arrival, and she received a courteous but disappointing reply regarding her audience with the president: His Excellency was ill but would take up the matter immediately once his doctors allowed him to resume engagements.

In Monrovia she made useful friends among members of the Liberian Frontier Force. Its battalions patrolling the interior of the country helped to keep peace among the tribes whose presence was far older than the USA's black settlers. From there, she was to travel a challenging 500 miles inland, but first she needed to arrange her guide and carriers and finalise her paperwork.

However, she had arrived at a hectic time for officialdom, because Liberia and the USA were in the throes of negotiating a major and historic deal, the Firestone Concession, which would give the USA the right to obtain rubber in return for its protection against colonial neighbours who were eager to annexe the tiny republic. All the people she needed to see were busy and despite numerous promises, her departure kept being delayed: 'I had said goodbye to everyone at least six times, and I began to feel like the farewell performance of a prima donna.'

When she did manage to assemble a crew, she found on the morning of departure that they had vanished. In desperation she borrowed a car from one friend, and driven by another, and they dashed 15 miles to Careysburg, a town founded by the first US Baptist Missionary in Africa. There, in her frustration, she shamelessly poached a 'small army' of seventeen local men who were booked to work for a party of American missionaries and offered them greater rewards than they had been promised.

Disconcertingly, she found the reason for the disappearance of her original team was fear of 'the Devil Bush', a sort of 'Masonic society'. At that time of year they held more sessions than usual, and should a stranger accidentally or deliberately lay eyes on the 'Devil' or High Priest or his acolytes, the penalty was said to be an instant and horrendous death. Even Europeans who knew something of the 'Devils' warned Dolly to hide herself and advised her not to go up country yet. Naturally, she ignored them.

One character she would remember fondly from all her travels was her guide and interpreter in Liberia, whose physical appearance was so bizarre that it fascinated her – so much so that she would have hired him even without his excellent references from his former American employer. A short man, he had a large pendulous lower lip that hung down from his teeth and lips, six fingers on each hand and six toes on each foot. His eyes she never saw, because in the light he wore large blue glasses without which he could only see in the dark. She never knew what his real name was, but it sounded like 'Teacup', so she called him that, which satisfied him. He swore he feared nothing and promised to protect Dolly from all the devils in Liberia. In return, she promised him rewards if he got her back in one piece. She soon discovered that whatever he lacked in good looks he made up for in virility, and he already had a toddler daughter.

In addition to Teacup, who to her amusement would never use one word if several would do, she hired two other personal servants. The route, mostly through dense forest, was very difficult to negotiate and she was carried in a hammock when conditions allowed. For that and the rest of the load she requisitioned carriers from the chief of each town, at a rate of 1s per head per day and their food. Other than her personal baggage, there were tents and bedding, tinned food and general supplies, as well as gifts for people they encountered, to please and, if necessary, appease.

At the beginning, the local commissioner accompanied her for three hours to the next village. This had its advantages, for with an official behind the long train of carriers, there was none of the 'quarrelling, lagging, shifting and re-arranging of loads' to which Dolly was painfully accustomed.

For once, not only was the weather comfortable but the carriers who bore her did so 'smoothly, almost tenderly', without the usual excuses to rest by the wayside 'or drink or bathe in every muddy creek'. However, she had, most unusually for her, a moment of what she called 'primitive atavistic snobbery'. Carried aloft, she briefly forgot she was 'an unimportant cog in the great mechanical device of the civilised world' and, as she dozed, felt she had gone back 1,000 years and was travelling as kings once did; she had fifteen men 'to transport by the sweat of their bodies, one small white human and her necessities of life'. When she came to, she felt ashamed of 'such an unsuspected streak of barbarism'.

For the first two weeks, they travelled through deep and claustrophobic forest, 'dense walls of morbid green that seemed to imprison and wear one down', and sometimes went for hours without seeing the sky, so it was refreshing when they reached more open country which occasionally offered a view. On the outskirts of most of the larger villages was a 'Devil's Bush', whose barricaded privacy left her wondering what lay behind.

As always, she paid close attention to the tribes they encountered, such as the Pessis, Mandingoes and Manos, whose methods of adornment favoured by both sexes were the safety pin, used in their hair and in jewellery, and the cigarette card, usually representing English or American music hall actresses. A hat was a popular accessory, and in one photograph Dolly took, a Mano king wears a bowler hat as he poses with his people. She tended to be critical of those do-gooders, usually missionaries, who foisted upon native people their Western ideas of civilisation, especially in dress, for she felt it compromised their dignity.

Her arrival at Sanoquellah,[10] the largest town in the north-east of the country, was significant. Not only was it the most northerly point she reached, close to the border with French Guinea, but she was aware it made her the first woman to cross Liberia to one of its remotest points. She was also told that she was the first white person of either sex that many had ever seen. Soon she discovered there was an interesting group in the area.

Tired of tinned food and stringy chicken, she told Teacup she wanted to buy fresh meat. He told her she could not buy beef because the people of that area did not eat it. Instead, 'They eat man'. She was not altogether surprised, for a few days earlier she had heard rumours and references to alleged cannibalism ascribed to certain tribes. Four years earlier, Teacup had worked there as a messenger for the local commissioner and told Dolly he had come across members of the Leopard Society, whom the government were trying to stamp out for that very practice.

Dolly knew from historical accounts that cannibalism had been practised throughout the hinterland of West Africa, including the Guinea coast, but overt cannibalism had been eradicated in the more accessible and opened-up parts of Liberia – at least, that was the official line, for it was a sensitive subject. However, her own experience in other countries showed her that things sometimes went on in ceremonies that would not be permitted if they were known about officially. Three years earlier, she had heard stories from French 'negroes' in North Africa of the 'human leopards' of the hinterland of Sierra Leone and Liberia, men who wore claws on their hands and were reputed to rip up the bodies of their victims.

Her informants gave her the impression that it was not a practice that had its roots in superstition: they told her that the sole *raison d'être* of the Leopard Society was 'the gratification of a morbid taste in human flesh, or as a witty member of the Liberian Cabinet with whom I discussed the matter later at Monrovia expressed it, "perverted epicureanism!"' However, on her return journey, she would learn that in Monrovia's prison were two men accused of having eaten the hearts of small children, because their medical advisers had said it would improve their health. It contributed to her leaning towards the theory that, rather than having a gratuitous desire for human flesh, the organisation *was* originally based on superstitious beliefs or on some form of witchcraft that time had modified or obliterated, yet which had survived 'as a measure and method of protective secrecy'. She acknowledged her theory was conjecture rather than knowledge.

Several district commissioners told her that the Leopard Society still existed in the area and she was given details of those who were in prison serving life sentences. Further on in her journey, she was taken to see the prisoners, making her very probably the first European, and surely the first woman, ever to do so; she was even permitted to photograph them.

The membership of the society was explained to her. A man who was a 'Leopard' would not tell anyone, even his wife, unless she was one too; usually, he led the life of 'an ordinary peaceable townsman'. The leader would appoint times where they had to fast in the remote bush, when they dressed in a prescribed manner and imitated the gait and noises of a leopard. They then sought a victim – an unwary traveller returning home late, a woman washing clothes in a stream – and killed them (Dolly was not told if there was a special rite or method), then ate them, sometimes a few days later. They tended to confine themselves to their own tribe. Teacup assured her that she was safe, because 'they fear white man too much. But if you were already dead, I don't think they'd waste you'.

Dolly only twice met anyone who confessed to having eaten human flesh 'and both of them spoke of the present prohibition with regret and utterly without shame or real consciousness of wrongdoing'. One told her that most tribes did not care for the flesh of a woman as it tasted bitter, while the flesh of a white person or educated black one was 'acrid and unpalatable' because of the amount of salt they took with their food. What concerned her, from what she was told, was that the Leopards seemed not to eat the meat of any other animal or bird, even though it was plentiful.

She also learnt that if victims were not forthcoming for the Leopard Men, they resorted to magic. The legend went that when they had selected a person, a member would go outside the village at night and blow some 'medicine' from a pipe, apparently a type of snuff, while calling the victim's name. Even though the person might be asleep, they would unaccountably wake and go out and no one would be able to stop them.

Dolly asked two people for their views, a Liberian commissioner and a Baptist mission helper. The former said he neither believed nor disbelieved but said there were many things in life that one could not understand. The latter recounted a story from his own experience of a wife who sought the mission's help to save her husband from such apparent magic, which seemed to work until she woke the next morning to find her husband gone once more. She never saw him again.

Various methods were used to catch suspected Leopards, one of which had been used for centuries, whereby they were forced to drink bowls of 'medicine' made from poisonous sasswood bark. The guilty were said to die in agony within minutes, the innocent, after severe pain, vomited the poison and recovered. Dolly was incredulous that such a method might be used, although 'its efficacy seems to be admitted by quite serious-minded and knowledgeable persons'. One theory, she said, was that their guilt was pretty much known by the authorities to start with; another was that the guilty person's fear of being found out made their stomach produce secretions that reacted with the poison. Sometimes, spying was used by a reformed member to reveal a Leopard in return for a pardon. There were even 'smellers-out', who seemingly had the power of divination and who, when placed among suspects, would sniff and point out the Leopard man or woman. Dolly met one and asked how he did it. 'I smell. I know,' he told her with a bored air. There were several tribes in northern Liberia who were associated with cannibalism, one of whom were the Mano, but she found them cheery and hardworking and found it hard to believe they could be associated with such practices.

When it came to magic, she had earlier had her own disconcerting experience on approaching Sanoquellah. It was about 11 a.m. on a brilliantly sunny morning with a blue sky. 'Even the forest seemed hushed and somnolent in the withering heat' as they made their way along silently, she thinking of nothing in particular, when they were startled by an almost explosive sound, 'a long rattling roar of thunder that, rising from the right, seemed to roll over us like an avalanche'. They all looked up but still there was no vestige of cloud anywhere. It happened again, then twice more at intervals of about two minutes. Teacup offered an explanation: the men said there was 'a big medicine man' making the thunder. While Dolly thought that ludicrous, she could not think of a plausible explanation. The discomforting memory would remain with her.

Louise Walpole at Weyborne with her children, Dorothy Rachel Melissa Walpole ('Dolly') and Horatio Corbin Walpole, no later than 1893. (Courtesy Laurel, Lady Walpole)

Dolly at Mannington Hall with (presumed to be) Mr Purdy, the land agent, and other staff. (Courtesy Laurel, Lady Walpole)

Mannington Hall, Norfolk. (Author's photo)

Dolly with her mother, now Countess of Orford, and Joey, the family's Yorkshire terrier, c. 1897. (Courtesy Anthony Palmer)

Wolterton Hall, Norfolk. (Courtesy Anthony Palmer)

Wild swimming. (Family papers, Norwich Archives)

'We were three Lady Dorothys together at Wolterton', 1907: (from left to right) Lady Dorothy Nevill; her niece, Dorothy, Duchessa del Balzo; and her great-niece, Lady Dorothy Walpole. (Courtesy Laurel, Lady Walpole)

Bust of Robert Horace Walpole, 5th Earl of Orford, (left) and Dolly. It is likely these busts are those he commissioned when they were in Florence in 1912. (Author's photo, taken at Wolterton, November 2021)

The Earl of Orford and his second wife, Emily Gladys Oakes, on their wedding day, 15 September 1917. (Courtesy Anthony Palmer)

Lady Anne Walpole at Wolterton, *c.* 1925. (Courtesy Anthony Palmer)

AN IMPENDING WAR WEDDING
A New Portrait of the Bride-elect.

Photograph by Yevonde—Inset by Hoppe

LADY DOROTHY WALPOLE

The only daughter of Lord and Lady Orford, who is to be married to Captain Arthur Mills, Duke of Cornwall's Light Infantry, whose portrait is inset, at St. Paul's, Knightsbridge, on June 22. Lord Orford, who is the fifth of his line, succeeded his uncle, the late earl, in 1894, his father having been the late Hon. Frederick Walpole. Lord Orford was formerly in the navy and attained to the rank of sub-lieutenant, and is now in his sixty-third year. Lady Orford was a Miss Corbin, the daughter of Mr. D. C. Corbin of New York

Announcement of the engagement of Lady Dorothy Walpole with inset of Captain Arthur Mills, from *The Tatler*, 21 June 1916. (© Yevonde Portrait Archive/ILN/Mary Evans Picture Library)

LADY DOROTHY MILLS

Who is shortly leaving on an expedition to the middle of the Sahara—main objective, Timbuctoo. Lady Dorothy Mills is well known as a clever novelist, and is the author of, amongst other novels, "Card Houses." A few years ago Lady Dorothy Mills went on an expedition to Biskra and the country 150 miles south of it

The cover of *The Tatler*, 10 January 1923, before Dolly left for Timbuktu. (© *Illustrated London News*/Mary Evans Picture Library)

On a camel in the Sahara Desert, 1920s. (Provided by Dolly for *More Heroes of Modern Adventure*)

Touaregs in Timbuktu. (By Dolly, from *The Road to Timbuktu*)

Lord Balfour opening the Hebrew University in Tel Aviv, 1925. (From *Beyond the Bosphorus*)

'My start from Careysburg'. (Provided by Dolly for *Through Liberia*, Schomburg Center for Research in Black Culture, Jean Blackwell Hutson Research and Reference Division, The New York Public Library Digital Collections, 1927)

Dolly's interpreter in Liberia, 'Teacup' (right), with Vani, another of her team. (By Dolly, from *Through Liberia*, Schomburg Center for Research in Black Culture, Jean Blackwell Hutson Research and Reference Division, The New York Public Library Digital Collections, 1927)

Dolly in travel clothes. (Photo by Lassalle, Baker St, provided by Dolly for *The Golden Land*)

'Leopard' prisoners at Tappi, Liberia. (By Dolly, from *Through Liberia*, Schomburg Center for Research in Black Culture, Jean Blackwell Hutson Research and Reference Division, The New York Public Library Digital Collections, 1927)

Liberian devil mask and doll. (By Dolly, from *Through Liberia*, Schomburg Center for Research in Black Culture, Jean Blackwell Hutson Research and Reference Division, The New York Public Library Digital Collections, 1927)

Dolly at Sanoquellah, Liberia. (Provided by Dolly for *Through Liberia*, Schomburg Center for Research in Black Culture, Jean Blackwell Hutson Research and Reference Division, The New York Public Library Digital Collections, 1927)

In a rest house at Labé, Guinea. (Provided by Dolly for *The Golden Land*)

Leaving Boké, Guinea. (By Dolly, from *The Golden Land*)

Crossing a river by ferry. (By Dolly, from *The Golden Land*)

Balanta men. (By Dolly, from *The Golden Land*)

Dolly after returning from Guinea, *The Tatler*, 6 May 1931. (© *Illustrated London News*/Mary Evans Picture Library)

Foulah woman, Guinea. (By Dolly, from *The Golden Land*)

Goahiro Indians. (By Dolly, from *The Country of the Orinoco*)

By the river in Venezuela. (Provided by Dolly for *The Country of the Orinoco*)

Women of the Motilones, Venezuela. (By Dolly, from *The Country of the Orinoco*)

At Vevey on Lake Geneva with the exiled Countess Catherine Károlyi and her son, Adam Károlyi, 1932. (Courtesy Angus Sladen)

At Vevey with 'Freddy' Sladen and Bobby Howard. (Courtesy Angus Sladen)

Dolly, *c.* 1932. (Courtesy Angus Sladen)

Dolly and her nephew, Anthony, in the Stone Garden, Rosemoor, 1952. (Courtesy Anthony Palmer)

9

BLOODY JACKALS

As much of her journey necessitated her being carried in the hammock, often for eight or nine hours a day, it felt like 'passive endurance', and now that they were deep into the interior, travelling was much less comfortable than it had been earlier. The heat had become oppressive, making it impossible to take any exercise except in the early morning or evening. They waded through foetid swamps, where unseen birds and animals fled from the sound of their footsteps and 'deathly cold and supple snakes, slim ones, minute ones, slid away at the sound of one's coming'.[1] Driver ants presented a terrible danger. They were deadly insects that ate their way through anything in their path, from leather boots to human bones, as the remains of a missionary had attested a few years earlier.

After Tappi (now Tapeta), where she was allowed to photograph the Leopard Men prisoners, they continued to Gblor, and so began the week she would describe as the longest in her life, 'so packed with troubles of every kind, troubles that seemed trivial as sprats in the recollection, but that loomed large as sea-serpents at the time'.[2] Her team had changed several times, with the arrangements supervised by Teacup, but now, for the first time, it comprised mostly women, who were excellent carriers but some seemed so young that Dolly felt it was almost child cruelty. No sympathy was offered by any of the women when a girl of about 14 collapsed, clutching her sides and whimpering. She herself seemed surprised by Dolly's attention, and after drinking a cup of her tinned milk she recovered and took herself off home in the forest.

At Gblor, Dolly came across the notorious 'gri-gri bush' she had heard so much about, 'that mysterious villegiatura[3] that officially turns the African maiden from a child into a woman'. Seven young girls aged 12 to 16, whom she described ominously as 'virgins', were segregated from the others and dressed scantily alike, their hair 'newly and elaborately tortured into a multitude of tiny plaits decorated with shells and metal skewers and their round baby faces freshly painted in the most grotesque pattern of livid white'. Here, even the 'Devil' of the Devil Bush gave travellers every chance to avoid seeing him by ensuring the road ahead was clear before venturing out.

Later, in the village, in the absence of its king, she was welcomed by his head wife, a handsome woman of about 45, wearing the usual scanty clothing but also magnificent jewellery — huge stone beads interspersed with leopards' teeth, 100 copper rings on her fingers, and her arms from wrist to elbow and legs from knee to ankle entirely covered with heavy bracelets and anklets of beaten brass. She presented Dolly with a large bottle of palm oil and received in return a bottle of pomade, mirror and comb, which she accepted 'with dignity, not unmixed with feminine thrill'.

Through Teacup, the queen asked Dolly questions — the usual ones about her age and marital status, and whether her father was a big chief. She had to think quickly, reckoning that a British peer of the realm was probably equivalent to a chief of quite decent dimensions. When she asked Dolly how big his kingdom was, she could not remember the number of acres 'of pheasant coverts and broad wheat fields' of her childhood home in Norfolk nor what it would mean to a tribal queen, so she answered in terms of the distance to a far-off Liberian town, which seemed to impress her.

Frequently, though, the harsh reality of women's lives was seldom far away. That evening in the compound where she was staying, she was aware of an altercation between a Liberian messenger and a young soldier from different tribes who did not understand each other's language, and it introduced Dolly to the grim custom of 'pawning'. The soldier was hard up and had previously pawned his woman to the messenger for 10s and had now come to redeem her. The messenger wanted to keep her and it seemed she was fond of him. He claimed the garment she was wearing – a strip of cloth which Dolly thought could not have cost more than 10 or 15*s* – was worth £4 and he would not part with the girl until the soldier paid him.

Although slavery was forbidden by law in Liberia, pawning was still allowed. Dolly discovered later that 'departmental regulations' allowed a man to pawn any relation or dependent of his for any price to anyone except a foreigner. It had to be with the pawn's consent, in the presence of the chief, and be accompanied by a token by way of identifying the pawn, such as an item of jewellery: without a token, the transaction was deemed to be slave dealing, the fine difference between the two of which escaped Dolly. There were other rules too, largely favouring the holder of the pawn. For example, if the pawn was a woman and bore the holder a child, he could keep it, provided he compensated the woman's real owner.

Dolly never heard how the matter between the two men was resolved but the argument continued into the night. Next morning, as she waited for the carriers to arrive so they could leave Gblor and head south, the world 'seemed full of love and finance', for it was Teacup's turn for trouble. A group of townspeople sought Dolly out and told her of his disgrace. He had spent the night with a young woman who was promised to another man, who was now demanding his legal due – £3 – for his damaged trust. It meant Teacup could be forcibly detained until the chief returned from Sanoquellah, leaving Dolly, disastrously, without interpreter or transport supervisor. She decided to try to sort the matter out herself.

The young lady, who she estimated to be 16, was brought before her and seemed 'quite unabashed' by it all, showing more interest in Dolly than anything else and scrutinising every detail of her person and clothing. Using a mix of basic legal and practical arguments, 'among a din that sounded like that of a parrots' house', Dolly eventually managed to negotiate the sum down to £1, for which she handed over 15*s* and left Teacup to make up the rest. Afterwards, she looked at him, 'the Don Juan [and] couldn't help speculating on the well-known but ever surprising fascination that perfectly hideous men so often have for our sex'.

The episode, though not without its humour, also brought to Dolly's attention the fines system that the Liberian authorities used in the larger towns to enforce public morality with laws based on the old tribal customs of the country. While £3 was the fine for possession of a woman,

£12 was the amount for the seduction of a virgin, rising to £20 if the girl was not of marriageable age. The fine for abortion was £12.

Dolly's view of missionaries is seen in a warning against an over-zealous sectarianism which, 'while spreading the grace of the Christian God with the one hand, would hand out one of the Devil's worse scourges with the other'. This was in relation to the 'problem' of polygamy, where she heard of a chief who was refused admission to a mission in eastern Liberia unless he got rid of sixteen of his seventeen wives. Her view was that it was impossible to discard existing wives in a country where a woman had little standing and no means of self-support, for a man had a duty to support his wives. If polygamy were abolished, argued Dolly, 'there would be a number of women without resources who would be thrown back upon prostitution, casual or symptomatic, an evil that primitive West Africa has never experienced'.

After Gblor came unpleasantness. In contrast to the Gios tribe, whom Dolly had encountered earlier, 'cheery, musical folk', here were the Quanoh, 'an unprepossessing and surly people who seemed to take life grudgingly', and in whose region members of the Leopard Society dwelt and where (as villagers told her) the eating of human flesh was not always confined to the Leopards.

Then, just as they were staying 'in a singularly ill-adapted village', she went down with three days of malaria, which 'knocked everything but will-power out of me for some time'. She had never had it before and knew little of the treatment, and with no medication or decent food available, she lay aching and shivering on what seemed the hardest camp bed in the world, where she longed weakly 'for cleanliness and quiet and freedom from vermin, and cool drinks and palatable food'. Her team were worried, as one of them told her, not least because if she died, she might be eaten by the 'bad people'.

Unexpectedly and touchingly, an atypical Quanoh, who was a soldier during working hours, took a particular interest in her, and when she was able to climb into a hammock slung under the thatch of her hut, he visited her with his small son. An 'exceptionally good-looking' young man of about 25, dressed in a red loincloth and a few leopards' teeth, he spoke a little English in which he expressed sympathy for her and played soft music on a tiny instrument, saying he liked her and that she was good. She was still wobbly when they left, and her new friend accompanied them to the Nuon River, ensuring they carried her hammock carefully at the shallow crossing place, where he shook both her hands and told her he was sorry she was going.

However, there was no let-up in the discomfort. As they tried to recruit a fresh team of carriers from the reluctant villagers at Bukor, the weather broke with heavy thunderstorms that 'made the atmosphere dead and lifeless, seeming to lie on one like a dead weight, making one's head feel as if it were going to crack'. It was vital to push on to the next village that night, about 5 miles away, because a long uninhabited stretch lay between it and Bharzon, an important Liberian military post, where a degree of comfort and facilities awaited them.

Her temperature was now nearly normal, although her legs still felt shaky, and the women they had recruited were threatening to run away. Dolly was relieved when three young Frontier Force soldiers on leave volunteered to carry her for a head of tobacco. It was the first time she had travelled through the forest at night and except for the pinpoint of light made by the lantern in front, the forest 'loomed all round us above and below like a live thing [...] a being with great dark hands that might any moment drop down, extinguishing our tiny beacon'. The journey had a nightmarish quality. 'The very air, humid and stagnant, seemed alive with unseen things, the wet leaves of the marsh plants swept our faces with clammy, viscid fingers, the black ground squelched under our feet with the sucking sound of oozing water.' A multitude of sounds, animal and bird, reverberated in the pitch darkness. 'It seemed endless, that trek through wet swamp and forest and exhausting to me at any rate, whose teeth were chattering and whose bones no longer seemed able to hang together.' It was with relief that, only half conscious, she 'sank down in a little bat- and rat-haunted hole of darkness, called a hut' in front of a fire that one of her men, Vani, had run on ahead to light.

Next morning, they needed to set off in good time for Bharzon before the weather deteriorated but some of her men had disappeared, so she started off ahead with one of her lead men, Meyer, and the hammock carriers, leaving Teacup to come on later with the rest of the team and the loads. However, the hammock carriers started a mutiny, and after half an hour's altercation, during which her only ally was a chief's daughter, one of them started shouting insults at her and 'threw his end down with a vicious bump, half braining me with the pole'.

To Dolly's frustration, two hours of precious daylight had already been wasted and, still feeling fragile, she completely lost her temper. 'Infuriated with the man's impertinence and by the pain in my head, I hurtled out of my hammock, and went for him like a maddened terrier with my sun umbrella.' It broke, but he was surprised by the onslaught and ducked and turned and Dolly kicked him up the bottom with her nailed boot, which delighted Meyer, whereupon the man 'docilely picked up the hammock and gave no trouble for the rest of the day'. Then the chief's daughter got involved, shouting at the men with arms akimbo, although Dolly could understand only a little of her ten-minute tirade: '"A sick woman, and small, and you refuse to carry her!"'

But things were set to become still more miserable as they set out on the last part of the journey to Bharzon, as a heavy storm broke, bringing 'pitiless rain' as heavy as hail. It 'sluiced off the bare backs of the men like a powerful shower-bath' while trees around them were struck by lightning and creeks were turned into tiny torrents. The fragile wooden bridges were broken, and she had to get out at each one and wade across. How blissful it was to reach Bharzon that evening, where they presented their 'clammy selves' at the barracks of the Liberian Frontier Force, and to hear the welcoming voice of an officer she had met earlier. He had prepared two clean rooms for her carpeted with mats, where she and Meyer ascertained how much of the load they still had left and tried to dry it out.

The hospitality did not extend to an actual bed, however, so she had to borrow one, nor to food, so she and Meyer had to eat the soggy remains of their supplies while she daydreamed about hot baths and succulent dinners. He was looking forward to a mound of dried fish he had saved for breakfast, and Dolly knew she behaved 'meanly and unchristianly' when its smell made her move it away in the middle of the night, despite her knowing that rats shared the rooms.

At last, the longest week of her life was at an end and so – almost – was her expedition. Before then, however, there were still tortuous stretches of forest to endure on their way to Nyaake, by which time even Teacup had stopped his giggling. How refreshing it was to leave the world of green behind and smell the salt tang of the Atlantic Ocean as they reached Harper, where her team were agog at the prospect of port life and she looked ahead to the 'great world of ships and trains, and quick, feverish life of work and play, of pleasure and responsibilities, a world that I had forgotten'.

She eventually met the president, who gave a ball in her honour. In Harper was a sight that always irritated her: people dressed in 'an assorted jumble of European clothing – and again I cursed the pale gods of civilisation who brought reach-me-down clothes to Africa!' She herself looked a mess, her hair still in the tight little plaits she wore when travelling – she never had it 'shingled' in the English fashion – and clad in 'disreputable shorts [...] a torn and jaded shirt and gaping boots', so much so that a large crowd of jeering African children followed her as she headed for a store run by a British company, Messrs Woodin, for which she fortunately had a letter of recommendation and they expertly kitted her out for her return journey.

By contrast, shortly after she arrived back in England on 19 April, a journalist who visited her at Ebury Street found it hard to believe she had been travelling in the heart of Liberia clad in shorts and shirt, for here she was elegantly attired in a beige jumper suit, lizard-skin shoes, silk stockings and long amber necklace.[4] As usual, she had brought back with her a number of trophies, including a devil mask and a sort of club with which she posed for pictures. She had written her next book already, she told the reporter, much of it done in her tent during three days of rain and needing to be dried out as the pages had stuck together. But there was to be little civilised chat about her journey. As soon as she mentioned having been to cannibal country the press began circling.

Sensational headlines appeared. 'Lady Dorothy Mills returns from Cannibal-land unbeaten – and uneaten,' trumpeted an American paper. Some British titles also emphasised the cannibal aspect. 'An Earl's daughter among Cannibals: Orgies of Human Leopards: Lady D Mills tells of Cannibal menus,' shouted the *Daily Express* in a headline of several inches, squeezing in 'Wives £12 each' underneath.[5]

However, lurid headlines were one thing but misrepresenting facts that Dolly had taken pains to point out in the article itself – that she had come across evidence of cannibalism only in the deepest hinterland and not on the coast, and such people were not representative of Liberians generally – was quite another, and unleashed an extraordinary backlash. A few days after her return, she received a terse letter from Cornelius Dresselhuys, the Liberian Minister in London, criticising the interview that had been published in the *Evening Standard*. He said she had given the reporter to understand 'that Liberia is "a country peopled entirely by cannibals". I, with others who have travelled extensively in Liberia, can however assure you that cannibalism does not exist there today.' His next and final paragraph must have incensed her: 'I should be glad if on future occasions you could omit these indelicate and false criticisms as they can only create unnecessary annoyance to the people whose hospitality you enjoyed'.[6]

Dolly's response was swift, controlled and to the point. 'I am surprised and very much regret that you have judged me so hastily by those criminally incorrect reports,' she wrote, for surely he must be aware 'how conscienceless newspapers are when trying to produce a "striking" story'.[7] Her husband could corroborate that she had asked her interviewers to differentiate between the coastal Liberians, whose hospitality and assistance she had described glowingly, 'and the bush natives of the hinterland'. It was not the journalists she blamed so much as the subeditors, 'who blue-pencil what does not interest them', and she was intending to write to all those who were culpable with a view to seeking an apology. She said that when her new book was published, she would send the first copy to the president and other officials, the contents of which would vindicate her.

In a paragraph that would surely annoy Mr Dresselhuys, but which could not be left unsaid, she told him that 'cannibalism as represented by the Human Leopard Society does still exist in a large portion of the Liberian Hinterland, the eastern half recently subdued and less settled, which I do not think you visited'. In addition, every Liberian commissioner with whom she stayed, 'whose names and posts I can give if required', had spoken to her about it, shown her proof 'and spoke of their difficulties in dealing with it'. If that were not enough, she had stayed in the same compounds as recently caught convicts and had later discussed the issue with the president and other officials. Finally, she invited him to ring her if he wanted to discuss it further.

She copied the correspondence to Mr Smallbones, the Chargé d'Affaires she had stayed with in Monrovia, with an expressive letter of her own. 'I am more distressed and disgusted than I can say at the monstrous and untrue turn our newspapers have given to my accounts of Liberia,'[8] she wrote, and asked him, should the subject arise in Monrovia, to give her letter the necessary publicity. She conveyed how much emphasis she had given to the lack of connection between 'the Americo-Liberians and the aboriginal natives of the upcountry bush', and listed other subjects she had covered in her interviews. 'All this was omitted in print and my brief account of the Human Leopard Society was seized on by a Press that battens on sensation. "Earl's daughter among canni-bals" is about the intellectual limit of present day journalism.' Anything negative that appeared to come from her, she assured him, should have been countered by her broadcast on the wireless a few days earlier and by several lengthy articles she had written.

On the same day Dolly sent him another letter, this time handwritten and informal. 'Dear Bones,' she addressed the diplomat with whom she had clearly established an easy relationship, 'I'm afraid you'll hate me for all this but it's not my fault.'[9] She freely expressed her distress at feeling she had let him and the Liberians down and her contempt for the press, to whom she had been 'giving hell' and who were 'just a lot of bloody jackals who care nothing for equity and everything for sensa-tion'. Admitting she was in a filthy temper, she ended, 'I hate England & its inhabitants and most of all its Press. The West African bush is a rest cure in comparison.'

Mr Smallbones came up trumps in his official response to her.[10] He had discussed the matter with the president who, while attempting to split hairs about the definition of cannibalism, admitted that he had had a number of men hanged for the practice, 'whatever its name', and also currently had in prison the two men accused of eating children's hearts. Bones wondered what Mr Dresselhuys would make of this in his denial of the existence of cannibalism in Liberia. Further, Dresselhuys' criticism that Dolly had somehow betrayed the hospitality she received implied that, in return, she was 'expected to act as Liberian propaganda agent'. That was nonsense, said Bones, because 'permission to visit the country can hardly be called hospitality and you paid for your own expenses'. Even the President's Ball given 'in her honour' was in return for Bones having invited ministers to Dolly's reception to make up for not inviting them to an earlier event when mourning for Queen Alexandra prevented it. He concluded, 'You are not their paid press agent and as long as you give truthful impressions in a friendly manner, they cannot complain.' Smallbones helped again in October by giving instructions for the distribution to the president and others of six early copies of *Through Liberia*.

That these exchanges were happening in confidence at the highest diplomatic levels meant they could not inform members of the public, who could chip in with their responses to articles in the press. From London, Robert Blay, a 25-year-old Ghanian and President of the Society of the Union of Students of African descent, wrote in protest about her statement that cannibalism existed in Liberia and also that she had described Liberian women as being 'beasts of burden' and of 'facile morality'.[11] Dolly responded (surely wearying of it by now) with her own experiences of the Leopard Society and of the women who had worked as her carriers while their men 'idled away the day'. As to their morality, she was simply pointing out 'the ease with which a wife can be purchased'.

She also found she had an adversary in another person she had never met. Leonard Leighton was a West African produce merchant in London, who wrote to the *Daily Mail* saying that in four years of living in Liberia, he had never heard of the 'drastic cannibalistic tendencies'[12] ascribed to them by Dolly, which forced her to respond publicly. But this was not the only gripe Leighton had. It was not Dolly but his wife, he claimed, who in 1908 was the first white woman to cross Liberia to the French frontier, which she had done as part of an expedition with him. He sent the paper a piece about it from the *Evening News* of 1910[13] after his wife, prompted by another article by a woman traveller, told that paper of her own experience. (She also said one of her carriers was eaten by cannibals, a detail Leighton seems to have overlooked.)

In responding, Dolly made no reference to the fact that clearly Mrs Leighton had neither travelled as a lone woman nor been responsible for organising and leading the expedition, for that was not the point here. Instead, she graciously acknowledged that it must be annoying for Mrs Leighton not to get the kudos she deserved but said that she must have reached the French frontier by a different (and shorter) route, for at the point Dolly reached, the indigenous people said she was the first white woman they had seen. She was also probably the first white person ever to have reached there. The *Daily Mail* noted that Mrs Leighton's route described in the *Evening News* supported Dolly's explanation.

Ironically, the Liberian Government's concern over Dolly's mention of cannibalism would pale into insignificance within four years, when in 1930 an indictment against Liberia was drawn by an international commission which comprised two black members, ex-President Arthur Barclay of Liberia and Dr Charles Johnson of the USA (to demonstrate that their findings were not based on racial prejudice), and an Englishman, Cuthbert Christie. It found that the American-Liberians in control of the government at Monrovia had shamefully exploited the natives of the hinterland and for private gain. The exploitation took several forms, one of which was pawning, the revelation of which shocked the world.[14] It would be gratifying to think that *Through Liberia* might have played a part in its exposure.

Certainly, Dolly's book was used as a source of information for the Harvard African Expedition of 1926–27, one of whose purposes was to make a medical survey of Liberia, which was less known from a medical standpoint than any other country in Africa. In a paper it presented afterwards, it referenced *Through Liberia* as a source. Although members of the Firestone Company had earlier provided useful information about Liberia generally, they had given only vague data about their route, whereas Dolly's route, while similar, was longer. Her book is also referenced in a modern study of violence against women in Liberia.[15]

On a lighter note, as the stories about Dolly were appearing in the press, Princess Elizabeth (the future Queen Elizabeth II) was born to the Duke and Duchess of York in Bruton Street, a few doors along from Dolly's late mother's house. *The Tatler* succeeded in mentioning both the royal birth and Dolly's cannibals in the same paragraph, being two of the three chief topics of conversation that week (the third concerning a new godson for George V).[16]

By the end of April, she needed something to take her mind off it all and what better way than to pay attention to Arthur. He was in the news too, for his latest novel *The Danger Game*, written after his first visit to the USA, was due to be published in June and was being serialised in advance by the *Daily Mirror*, amid much fanfare. 'Greatest Love Story of the Year' headed a column publicising the first instalment, accompanied by a handsome photo of him. 'With love as the prize, women will do many strange things. Mr Mills understands modern women and writes of them fearlessly and realistically.'[17]

His own 'modern woman', meanwhile, was absorbing the reviews of *Beyond the Bosphorus*, which also came out that year. They were consistently positive. *Truth* said Dolly was 'an acute observer, with the added gifts of insight and imagination'. Her description of Balfour's visit 'exhibits the Zionist movement in a light which will give those people something to think about who take their views of Zionism from the *Daily Mail*'.[18] 'Lady Dorothy Mills is as vivacious as she is courageous,' said the *Daily News*. 'She makes her very miseries amusing [...] and is observant and critical [...] This is undoubtedly one of the better sort of travel books.'[19]

In late July, Dolly and Arthur left for their annual holiday, this time on a European motoring tour with friends Algernon 'Algy' Sladen and his wife, Winifred, known as Freddy, who was an interesting character. Her father, John Dupuis Cobbold, was head of the Suffolk brewing dynasty and her mother, Lady Evelyn, a sister of the Earl of Dunmore. Bucking the family trend, Freddy had become a socialist, as well as a pipe-smoker and speaker of Serbian. From around 1919, she and Algy had been renting a house in Switzerland, which since the Great War had become a haven for refugees and revolutionaries; their friends included the exiled Hungarian socialist leader Count Mihál Károlyi and his wife, Katalin (Catherine). Freddy's views would get her into trouble, however, and give a glimpse of a rapidly changing Europe.[20]

The two couples started out from Belgium and by August were in Munich, where they separated, the Sladens going on to Austria, to finish in Italy. However, nothing had been heard from them since. On 5 October, Freddy's aunt in England, Mrs Reginald Coke, who Dolly also knew, received a letter from Freddy written in French headed 'Prison, Volosca'. Freddy said she had been arrested for speaking against the fascist leader Mussolini and was not allowed to write in anything but French. It transpired that, in September, she and Algy were in a café in Croatia with a Hungarian acquaintance, who mentioned an attempt a few months earlier by a woman (Violet Gibson) to kill Mussolini. He then started praising fascism, which angered Freddy, and they had a discussion. Later, he repeated their conversation to an Italian woman who also knew, but disliked, the Sladens, and she immediately denounced Freddy to the authorities.

Three weeks later, she was arrested and imprisoned at Volosca, where she was held for three days and nights before being released on bail, pending her trial in November. The media went wild about the imprisonment of a British subject, especially one who was the niece of an earl (as the newspapers kept saying), but for once, it may have done some good, because when he returned from England, the British Ambassador in Rome arranged for her immediate release. It was unfortunate that Mussolini's sensitivity was heightened by the earlier assassination attempt, but the episode sounded alarm bells generally and for a time the British press was full of cautionary tales and warnings to travellers.

When *Through Liberia* came out, it further enhanced Dolly's reputation as an explorer and writer. In a lengthy review, *The Scotsman* called it 'a work of remarkably varied interest [...] Above all [she is] a traveller with a sense of humour which will endear her to her readers.'[21] Another said that while there was little new information about Liberia:

> The general reader will find what goes to make a really good book – a lady persistent and courageous, facing the difficulties of travel in a land almost uncivilised; a reporter, quick to note and to set down vividly all the things that came new to her observation [...] The book has the driving power of sincerity; it is not merely written for a market.[22]

The Sketch said, 'Whatever the future development of Liberia [...] this record of a woman's exploration will remain a part of its history.'[23]

Dolly's views were often sought, including on a sensational story that broke in a British newspaper in October and was picked up enthusiastically by the US press. The Anglican Bishop Packenham-Walsh, working in a college in Calcutta, said he was told by the Reverend J. Singh that he had heard rumours from villagers of demons in a remote part of Bengal. Singh said that when he investigated, he found two girls aged 2 and 8 living in a wolves' lair and behaving in a feral way, having apparently been fostered by the pack.

The story would turn out to be more complex than first appeared, but initial responses were sought from the eminent zoologist, Professor Julian S. Huxley, who said it had echoes of Rudyard Kipling's Mowgli and that proper evidence was needed, and from Dolly, 'the distinguished explorer and writer'.[24] She said that in West Africa she had been told that small children were sometimes taken off and brought up by a big species of monkey common to the country. However, 'It does not seem probable that anything but a very small baby could be brought up by wolves in that way,' she said. 'One would think that when the child reached three or four years of age, its behaviour would arouse suspicion in the animal.' Much would also depend on the kind of food it ate. She concluded that 'the story seems just possible but very rare'.

In November, Dolly started giving lectures, the first one at a venue in Park Lane, London, where she amused her audience by saying that walking around Africa was much less terrifying than talking about it. Not everyone was impressed by her achievements – usually men. She received a newspaper cutting in the post of a short piece she had written about Africa, with a letter saying, 'Women like you make me sick' and signed by 'A disgusted naval officer'.[25] It seemed to Dolly that hers was a harmless little article and she wondered why he was so upset.

Again, her output astonishes. Just in time for Christmas that year, her novel *Phoenix* came out. A science-fiction story, it centres on a wealthy woman who mourns her lost youth. On a cruise she meets a Syrian scientist, who persuades her that she can reverse time by undergoing a surgical procedure he has developed. However, there is a catch: to have the operation, she must commit to him emotionally – but she loves someone else. It cannot end happily.

Reviews were mostly very positive, although one said that her idea was not original because it was similar to *Black Oxen*, a bestseller of 1923 by the American novelist Gertrude Atherton. However, Dolly might well have responded that the book evolved from her own vanity, for she admitted it was piqued when she was told in the bush that at her age, 37, she was considered well and truly past it. Besides, at least one review said the book struck a fresh note and in places was daring.

In what was a golden age of children's books, exemplified by authors such as Angela Brazil and Eleanor Farjeon, Dolly even wrote a story for *Mrs Strang's Annual for Girls*, a high-quality illustrated book published by Oxford University Press.

However strained her relations with her father, they were still in touch. She sent him an inscribed copy of *Phoenix*, possibly in response to his letter sent a few days earlier. Clearly, the book had made an impact in the earl's locality, although not positively. 'I am sorry I have shocked Aylsham!' she told him. 'It is true that in one chapter I gave a brief picture of "low high life" to act as foil to the virtue of the heroine and to point a general moral. The reviews that have so far come in are very encouraging.'[26]

She apologised for her note being hurried but she was 'simply rushed to death these days', and however good her intentions, she always seemed to leave things till the last minute. She and Arthur were about to go away. 'We have had a good let for the flat [in Ebury Street],' she told him. 'We reach Algiers next Sunday and motor down to a little oasis called Bou Saada, for Christmas and the New Year.' After that, she had no definite plans and would just wander after she finished a book she had started in the summer as the fancy or opportunity took her. 'Nothing sensational; I want a rest cure in the sun, after two or three very strenuous years of travelling, writing etc.' Promising she would write again from Bou Saada, she sent her love to him and good wishes to everyone at Wolterton.

Several Christmases were spent abroad, but the one at Bou Saada (meaning 'place of happiness') was probably one of the more civilised in terms of comfort, where she and Arthur were unlikely to have the sort of experience she had suffered in the Sahara of north-east Nigeria. Her guide had proved to be 'a false prophet and a guide in name only',[27] and took her to a well on which they had pinned their hopes, only to find it contained a dead jackal. Their best camel died suddenly, and they were engulfed in a sandstorm.

Dolly was hoping to spend Christmas at an outpost of the Foreign Legion, 'those poor gallant exiles who police the barren wastes of the Sahara', knowing they would have shared with her their 'tinned salmon, warm, sweet champagne and last year's gramophone records'. Instead, they got lost, and on the way to their night's shelter they came across vultures circling a dying man who had been left behind by another caravan.

When they reached their camping place at sundown, Dolly was so stiff from riding that she had to be lifted from her camel and for Christmas dinner had to be satisfied with the remains of the food in her saddlebags. 'It consisted of a tin of sardines, a handful of sand-coated dates, a small hunk of native, unleavened bread and a mug of water from which not even my last pinch of lemonade powder could eliminate the taste and odour of the goat-skin in which it had travelled.' Nevertheless, she was so hungry that it was more welcome than the finest dinner she had eaten in London.

On another Christmas Day, this time in the Sahara in Algeria, she had sat in the marketplace in the shade of a tamarisk tree, drinking innumerable cups of sickly sweet tea as she watched the bargaining and gossiping among 200–300 Arabs of different tribes. Their number was swelled by a big caravan that had arrived from the Niger the night before after crossing the Tanezrouft, known as the 'Land of Terror', the Sahara's driest and most desolate area. 'The air smelt aromatic and strange,' recalled Dolly, 'and rang with the loud laughter of black naked figures from the South who wore Pagan amulets round their necks and skewers of wood and metal through their broad noses.'[28]

As the sun went down and little fires sprang up in the marketplace, the sounds of revelry rose from the tiny cafés, 'the scream of flutes, the dull beat of tom-toms, the shrill laughter and clanking anklets of dancing girls', all which died down as the night wore on and 'the oasis slept in the moonlight – a brief and tragic sleep that was suddenly rent by the sound of trampling feet in the women's quarter, the shouting of men and the hysterical screams of women'. A beautiful dancing girl named Embarek had been murdered in her sleep, stabbed by her lover, who had found her in the arms of a rival, 'a handsome falconer from the Niger caravan'. For half an hour there was pandemonium, but then it suddenly died down. 'The lives of women are cheap in the Sahara, especially those of dancing girls,' Dolly noted grimly. So ended her Christmas that year.

Meanwhile, as she and Arthur relaxed at Bou Saada, the ghosts of Christmases past receded into the shadows.

10

THE OUTLIER

With the emergence of the 'modern girl' during the interwar period came a reassessment of the roles and expectations of the sexes. Dolly and her fellow explorer, Rosita Forbes, had 'explored ground that many men have not attempted, and endured hardships in some of their travels that would put an end to some young men about town', said one newspaper, and yet they remained 'essentially feminine and dainty'. Like Lady Montagu, the eighteenth-century explorer, 'they prove that a mere appearance of masculinity is no asset for the woman who really gets things done'.[1]

During 1927, Dolly wrote a feature for the *Daily Mirror* called 'The Spartan Woman of Today' (referring to the hardy women of Sparta in Ancient Greece), subheaded 'New Race of Feminine Die-Hards and their Wonderful Exploits'.[2] In it, she paid tribute to those women who were successfully competing with men in the more perilous sports and professions. Her article was prompted by a male friend – a 'he-man', as Dolly dubbed his type – who had spoken resentfully of a professional woman driver who had made a world record, calling her one of the 'new sex'. Dolly thought the expression was rather apt, for there was an increasing number of 'young Spartans, to whom their femininity seems to be a thing to be set aside and over-ruled. They have set themselves, and often very successfully, to do all the things hitherto considered impossible to women to out-do men in the fields of physical endurance.'

The first Women's World Games, held in Paris in 1922, had inspired regular international contests and women were excelling in many sports that had usually been the forte of men, such as rowing, running and football. As Dolly reminded the readers, they were also becoming known 'in aviation and motor-racing, exploration and big-game shooting'. To reach such a level of excellence, they had 'given all their best efforts of mind and body and endurance'. Further, they had to 'sacrifice most of the amenities of feminine life to their Spartan regime, have had to let love and marriage and motherhood go by the board', because society was not yet ready to enable them to pursue their goals without paying a price.

This type of woman champion was not unique to Europe and America, said Dolly, and she had found 'new' Turkish women who had fought alongside men during the war and recently, photos had emerged of the women's battalions in the new Chinese Army. An eminent woman doctor had told her it was 'quite a fiction that a woman's body is necessarily a fragile thing that requires cosseting and that with the present-day healthier mode of living there is no reason why she should not possess as fine a physique as a man'.

Dolly considered that the Spartan Woman was mostly made 'by economics and her new freedom'. Given the superfluity of women since the war, she said, many women did not want to be a financial burden on their fathers or brothers, so they worked, although they were often criticised for taking jobs away from men. However, in doing a man's job, a woman 'does not necessarily [...] make wage-earning harder for a man, since she relieves him of another mouth to fill'. There were also some women, and men too, who were simply not suited to family life. Given that favour was more often on the side of man, if a woman achieved success in some Spartan role, 'it is a proof that she is better fitted for it than the man who nurses a grievance because she has ousted him from it!'

The Spartan Woman, of a 'new sex', had interesting possibilities because she was 'a clever bit of evolution on the part of Nature, who cunningly adapts her materials to new necessities'. Should there come a time when the sexes were numerically more equal again, 'when domesticity is organised on a fairer and more co-operative basis, I think she will take the line of least resistance and resume her womanly role'. For the time being, she 'takes nothing from womanhood, she does no harm to the race'. The woman who was essentially (what was considered) 'feminine' would stick to her 'pre-destined role, while the Spartan woman sticks to hers, doing the things she does well, instead of making a probable mess of domesticity, leaving wifehood and motherhood to those better and more innately fitted for them'.

By suggesting that in time, Spartan Woman may revert to her traditional role, Dolly implies that she was, in part at least, adapting to cope with the current social situation, rather than preferring it. However, even Dolly could not envisage a world where combining motherhood with activities traditionally associated with men was a real possibility. As if to reassure any woman who might be worried about her role, Mr Justice Swift, a prominent High Court Judge, publicly said, 'Even in these days the profession of marriage and housekeeping is still a highly honourable one.'[3]

Some found it useful to distinguish between different 'types' of the new woman, perhaps to rationalise the phenomena. One writer[4] said women like Dolly and Rosita Forbes co-existed with 'masculine women' (such as some athletes), 'feminine men' and the 'Third Sex' (different from Dolly's loose term 'new sex'). This denoted a distinct sex between man and woman consisting of 'congenitally inverted individuals', a description which derived from a theory that attributed homosexuality to a failed correspondence between one's physiological sex and one's gender performance.[5]

For all the talk of women's prowess and toughness in partaking in traditional male activities, Dolly found the attitude of some of her own sex to be less than liberated. An example was the controversial issue of women continuing to take part in athletics in the Olympic Games at a time when the International Amateur Athletic Federation wanted to get rid of certain events as being too taxing for them, including the 800m race. Novelist Ethel Mannin said she was 'frankly bored by woman's persistent efforts to emulate the male' and there was no genuine desire to take part, rather a pose by those who were 'obsessed' in rivalling men in sports.[6] Lady Alexander said women should stop wearing 'those terrible running shorts' and concentrate on games like tennis and golf, because athletic sports 'enlarge our joints and make us unfeminine'.

Dolly's response was that 'the old-fashioned fetish that women cannot take part in athletic contests is dead. I think there are very few forms of athleticism that my sex cannot take up with distinction.' She considered the English girl to be graceful and agile when running and jumping and there was 'not a scrap of evidence to show that she suffers from such exercise'.

Gender fluidity and sexual experimentation were openly embraced in parts of society at the time. The 'Bright Young People' with whom Dolly was (unusually) pictured at Mrs Coke's party in the summer of 1927[7] included the fashion photographer Cecil Beaton in a dress and, in men's attire, photographer Olivia Wyndham, who would shortly leave her husband for her black female lover. Dolly wore a version of her explorer's outfit, with long shorts and accessorised luggage.

However, while people were coming to admire the independence and courage of female explorers, she still encountered hostility and, at best, condescension. 'It never fails to make me angry,' she wrote:

> … when after some experience on the fringes of the map that has entailed a measure of endurance or sporting spirit, the man who has shared or witnessed it thinks he is paying me the highest compliment in his power when he exclaims: 'What a sportswoman! One would never think you were a woman.'[8]

While she preferred to travel alone, if it came to 'the last ditch', she would rather have a woman than a man for a companion because while a man's physical strength 'might keep him alive a little longer', it would be the woman who 'remains to the horrid end', because she had more staying power.

She found many people were still 'inclined to regard the female wanderer with distinct suspicion, as a freakish creature whose sanity or truthfulness is at fault'.[9] She had lost count of the number of times she had learnt that she had never actually visited the countries she had written books about, or how many men of different races she was supposed to have committed 'indiscretions' with. At one dinner party, the man sitting next to her – neither of them knew the other's name – started talking about travelling, from which Dolly ascertained he had only got as far as Monte Carlo. He began 'a diatribe against women travellers, their morals, their sincerity, their achievements'. Mischievously she encouraged him to continue. '"Pictures in the papers […] nose-powdering and notoriety […] sex and sensationalism",' he went on, until she asked him on which women travellers he based his views. In the several names that spewed forth, hers was among them. A chance remark of the hostess conveyed Dolly's identity to him and for the rest of the dinner he 'looked like a boa constrictor who has swallowed a too big rabbit by mistake'.

By contrast, she also got letters from complete strangers that brought a lump to her throat, 'men and women who try to convey their appreciation and gratitude to one luckier than themselves who has been able to gratify the longing for the great spaces that is in them'. Sometimes, they wrote before she left on a journey to wish her good luck. On one occasion, a cleaner in a Scottish hotel who had read an article of Dolly's that had been left behind in a wastepaper basket sent her a faded bunch of violets and a prayer for her well-being. All this made Dolly feel it was worthwhile to continue writing 'to try and convey the wonders of the great world through the grey medium of printer's ink'.[10]

For Dolly and Arthur, much of 1927 was spent doing just that, she still contributing features on modern manners as well as travel and was much in demand as a speaker. She was considered 'one of the most entertaining storytellers in London […] with] an almost inexhaustible store of reminiscences, and many have a thrill as well as a comic touch'.[11] A digest of her Timbuktu book was also published. *Episodes from 'The Road to Timbuktu'* was one of several *Episodes From …* volumes in the *Readers of Today* series published by George Harrap & Co., which produced shorter and therefore cheaper editions of existing books.

After their summer sojourn at the beautiful and little-known Veulettes-sur-Mer in Normandy, Arthur's new adventure book, *The White Snake*, set in South America, was published after first being serialised in the *Daily Mirror*. As if they were in competition with each other (which some papers liked to say they were), Dolly's novel also came out.

Master! looked at the ever-topical issue of race, specifically mixed marriages, and what being of mixed race meant for individuals, both in emotional and practical terms. Her heroine is an artist of mixed race, part British, part Creole, who works in Paris and whose white lover says he would marry her if it were not for his sick wife. However, after her death, he does not do so. After giving birth to his son, she returns to England where she meets an African king who has been educated in Britain and is trying to reform his country. He takes her to Africa, where she feels she should belong, as she has black heritage, but she is never accepted.

Meanwhile, he falls prey to nationalistic elements, who resist his reforms, and is murdered. She feels she belongs neither to white culture nor black but considers her son is white. She turns down a marriage proposal from a white trader, who has been a good friend to her and her husband and returns to England.

The book also raised the topical and sensitive question of whether Africa could successfully be united with shared goals under a fictional 'League of Colour', which seems to have anticipated the League of Coloured Peoples started by Harold Moody four years later (although there were similar earlier organisations). Wanting her father to be aware of her latest work, she sent him a copy, inscribing it simply, 'To Papa, With Best Love from Dolly'.

The *Times Literary Supplement* said of *Master!*, 'Many of the African scenes and episodes in this book are picturesque and powerful.'[12] A specialist in African literature would later write:

> In contrast with novels which try to portray primitive life are those which have as their principal themes the nature and conduct of British colonial rule and which, while having a white bias, attempt to describe African social and political life and to achieve an African point of view.

He named *Master!* as one of the best examples of such books.[13]

It was sadly ironic that having paid tribute to women who took part in traditionally male activities, Dolly would find herself commemorating a pioneering relative who had gone missing in a flight. Princess Lowenstein Wertheim, born Lady Anne Savile, was the daughter of the

4th Earl of Mexborough and his second wife; his first wife was Dolly's great-aunt, Lady Rachel Nevill, whose premature death had left him with their young son, John.

A champion of aviation, Anne had financed a pilot friend's trans-atlantic flight on the *St Raphael*, a Fokker monoplane, with the aim of it being the first ocean flight westwards (to Ottawa), a more hazardous journey than west to east. Despite her half-brother John's protestations, Anne, aged 63, joined her friend and another experienced pilot as a passenger. They left England on 31 August 1927 in favourable conditions – and were never seen again.

In October, Dolly attended a memorial mass for the princess, at which buglers from the Royal Air Force played 'The Last Post' and the priest spoke of how, having been fearless in life, she would have been fearless in its ending. Although no trace of the plane or its occupants was ever found, it was believed they had crashed off the coast of Newfoundland. They would officially be declared dead in February 1928,[14] their tragedy highlighting the dangers of those exciting but still early days of aviation.

As 1928 began, with Dolly being photographed at the Wylye Valley Hunt Ball, looking the elegant epitome of 1920s style, it was hard to envisage her at home anywhere else. However, those who read her feature that month on black magic, in which she recounted some of her stranger experiences in the African bush, would be reminded of that other wildly different life.[15]

The year had started without Arthur, who had gone to find fresh writing material. He left Marseilles on a French steamer bound for Indo-China (Vietnam, Cambodia, Laos), where he would spend a month travelling, followed by a voyage to Hong Kong on a cargo boat and on to Hollywood to experience its heady glamour. His adventures would colour his next book, *The Blue Spider*, published later that year.

Dolly's novel *Jungle* came out, about a famous woman explorer who, having married a diplomat, finds his traditional ideas insufferable with tragic results. One review said it was an 'absorbing story with characters intensely alive'.[16] She was also writing features about her experience of black magic, particularly the widespread religion of animism, 'the belief in the power and universality of spirits'. She had come to real-ise that 'there is much we cannot understand in the primitive places of the earth, where humanity is very close and very receptive to the mysterious forces of Nature', and while she had been sceptical, she acknowledged that in some places, 'despite logic and common sense, my scepticism little by little drops away, for in Africa I am convinced the dark gods are not yet dead'.[17]

It would be a year of ups and downs for Dolly. In February, she went to Cannes for the rest of the winter, staying in the luxurious Cannes Palace Hotel, but fell ill. Arthur returned home before her in mid-March and felt compelled via The Times[18] to publicly deny that she had anthrax, which was rare in humans but highly infectious – the rumour seems to have stemmed from several newspaper stories from the Riviera.

What she was suffering from is unknown. The symptoms of anthrax can resemble those of primary syphilis, and while such a suggestion may seem distasteful, all it would have taken to become infected is a brief and singular encounter with a careless man – who could even have been her husband himself. Arthur, whose travels, although challenging, had culminated in his being invited on set to watch the beautiful and hugely successful Polish-born actress Pola Negri being filmed for a silent movie, claimed exhaustion and went to the country.

Dolly returned from Cannes in early April apparently recovered. In July, she was among a handful of distinguished travel veterans, including Lady Heath, the aviator and Commander Frank Worsley, captain of the Endurance, at a significant event in Kent attended by 7,000 people for the 'christening' of a Bluebird aeroplane, which had been bought for a worldwide youth charity. Another keen aviator, Mrs Ruth Knowles, was founder of the Friendship Fellowship, whose purpose was to give all young people, however poor, the chance to travel abroad and make friends. Aeroplanes, she said, were going to be one of the most powerful means of bringing the nations together in friendship.

Each guest gave a short speech, in which Dolly said that 'adventuring' was the 'greatest fun in the world' and she spoke movingly of how it had taught her that 'the most valuable and happiest asset in the world is the gift of making friends'. For her, they included a tribal chief in West Africa, who she said had saved her in a life-threatening situation.[19] She was also thrilled to have experienced her first flight, thanks to Lady Heath, the plane's 'godmother', who christened it the Friend Ship.

Dolly's novel *Phoenix* was entered for a major film plot competition, which was judged by a committee of film makers and critics, and out of 30,000 entries it made the shortlist of three. In September, along with Worsley and other eminent men, she was the only woman speaker at 'Explorers' Week' in London.

But she was reeling from a shock. Her father, aged 75, had decided for his health to live in New Zealand and had made Wolterton and other properties over to Bobby Vade-Walpole, aged 15, a distant cousin and son of the late Horatio, who had died in the war. In the absence of a son to continue the earldom after Orford's death, Bobby was the nearest male and would be the successor to the Walpole baronetcies. However, the estate was not entailed – it did not run with the titles, so there was no imperative for Orford's action. There was also no requirement to hand it over while he was still alive, but he did not want the worry of someone caretaking the properties in his absence abroad, having spent a good deal of time and money on their improvement and maintenance. He liked Bobby's mother, Dora, and perhaps felt sorry for the war widow and her son and elder daughter, Pamela.

The previous situation had been that under Orford's will, his wife Gladys would enjoy the properties for her lifetime, and after her death they would pass to Bobby. However, in March 1927, Orford was taken seriously ill from a carbuncle and diabetes and was in a nursing home for two months, which made him reassess the future. Discussions began in December that year, days after Dolly visited her beloved Wolterton; she was there again on Christmas Eve, seemingly oblivious to what was going on.

In a letter dated 16 December, Gladys told Dora that she and Orford had talked about the possibility of a transfer to Bobby sooner rather than later: 'I do not think I should much care about having the property & the house & most of the contents after his death.'[20] Perhaps Gladys was being magnanimous, for the 'property' included not only Wolterton, with its several thousand acres, but 'Waborne Hall & its Estate, Mannington & all the other houses on the Estate, with certain reservations as regards Mannington'. Alternatively, it may have been that Gladys felt she had no real choice in the matter and was presenting a united front to Dora, although it had ramifications – while Dolly had few expectations of her father, her half-sister Anne, now 10, would also lose out. To mitigate this, Orford declared that Anne should marry Bobby.

As Bobby was still a minor, Orford wanted the women to be two of the trustees. 'My husband is very anxious to keep the family together,' Gladys told Dora, 'and the interesting things which belong to an historical family in which we know you take a great interest.'[21] There would have to be a discussion about money and whether Dora was able to make a part payment for the transfer of the property. Certain restrictions would also need to be discussed for the ongoing benefit of the estates, such as the cutting of timber, and an entail put in place to protect future dealings. Once they had reached an agreement in principle, the lawyers could formalise it.

In August 1928, when everything was finalised and the new incumbents set to move in on 1 September, Orford, Gladys and Anne prepared to leave for New Zealand; his wife and daughter would eventually return to England to live at Rosemoor in Devon, which he had bought five years earlier as a fishing lodge and extended, and where they already spent three months every year. Orford had originally proposed to Dora that he continue to live at Wolterton while she lived at Waborne Hall, but she refused, preferring to move into Wolterton immediately from her home in Stevenage with 20-year-old Pamela (Bobby was away at Eton) and to let Waborne.

The homes Dolly had known since childhood and whose memories she still held dear had been ripped away from her. She knew then what she had always known in her heart, 'that I was not to be reinstated among my own kind; I came to know definitely, once and for all, that I was an Outlier from my tribe'.[22] On 1 August, Dolly returned briefly to Wolterton. 'I said goodbye to the past, for others were coming to take possession of my birthright.' Hour by hour she lived through her childhood, wandering through the woods and gardens where she had first heard the call of the drummer: 'Every path and corner had its little ghost to torment me [...] it seemed more beautiful even than I remembered it, more spacious yet more intimate, more really and truly a part of me.'

Dolly became the object of scrutiny. People came to see her, some of them 'simply to stare at the Outlier whose nose was out of joint, to see how she was taking it'. Others, though, who loved her, were supportive, 'some of them old now, and infirm, who had waited so long for me, and hoped so much from me, who gave vent to their grief and disappointment'. Her position was not helped by the fact that her father's action was inevitably widely publicised, the press surmising that it was probably the first time an estate had been handed over in the donor's lifetime to someone who was still a minor. One newspaper declared it 'one of the most curious transactions in the modern history of England's peerage',[23] while Bobby was proclaimed 'the most envied boy in England'.

If she knew about it, Dolly may have derived some bitter satisfaction from the fact that not everything was plain sailing for Wolterton's new residents. Less than a month after the Orfords left for New Zealand, a large fire broke out inside the chimney in the library where Dora and her daughter were entertaining friends and burned for five hours. In those early days, the earl could not help micromanaging the estates from abroad, writing often to Dora to check that nothing was being done that might jeopardise the land or upset long-standing arrangements with tenants or commercial shooting parties. But it could not offer comfort. Instead, Dolly did what came naturally to her and threw herself into preparation for her next expedition through French, then Portuguese, Guinea.

She anticipated it would take about four months and give her the opportunity to 'find a peephole through the barrier of mysticism'. She also wanted to add to her collection of tribal art and before leaving, she wrote to the Wellcome Museum in London in response to an advert about their acquiring such objects. Her intention, she told them, was to purchase 'specimens of nature (negro) art of all sorts', including carvings – 'fetish idols, masks etc.' – of which she hoped to give a small exhibition in London on her return. She invited them to contact her if they were interested.[24]

Her collection was already substantial. On the walls of the drawing room at Ebury Street hung half a dozen masks that 'from time immemorial have served as symbols of queer, rather dreadful age-old creeds and beliefs'. They included a devil-mask from the Liberian forest and 'a little formless image once given me by a disgruntled householder of Timbuktu, who had bought her for the luck her goodwill might bring him and she had promptly cursed him with a faithless wife and failing crops'.

In early December 1928, she sailed from Liverpool to Conakry, Guinea. 'To some countries one returns,' she wrote. 'West Africa is one of them.'[25] However, as she approached the coastline for the third time, she wondered why. Too often, it ruined 'one's clothes, one's nervous system, one's looks, and sometimes one's health, and one's morals'. It put obstacles in the way: inadequate boat services, lack of accommodation, 'health scares of malaria, yellow fever, sleeping sickness'. When she arrived on Christmas Eve, she felt exhausted 'more in mind than in body', and devitalised by cold weather and 'noise and so-called gaiety, utterly tired of the jargon of literature and the buzz of offices'. Although she intended eventually to go north to Portuguese Guinea, there seemed to be no information available as to how she might go about it, and she lacked the energy to think or make plans.

Soon she would be luxuriating once more in the familiar heat that curled through her veins 'like gentle fire' and smelling the healing aroma of vivid red earth and heavy green foliage. First, though, she had to deal with the hostile reception by Conakry's 'ambassador', an unshaven and ill-tempered French Police official, who demanded £40 from her under a new regulation she knew nothing about. She explained that once she had paid her on-board bill, she would not have enough money until she could get to the bank in Conakry. He said if she could not pay at once, he would refuse her permission to land, which meant she could be sent sailing indefinitely down the African coast. If she did try to stay in the town without visible means of support, the government could ship her home as a 'DBS' (a Distressed British Seaman). Fortunately, she was able to turn to her trusty shipping company, Elder Dempster, who sorted matters out for her in time for her to see something of the little tropical town before Christmas Day.

Never fond of Christmas, she was looking forward to spending it alone, as she so often enjoyed doing; it was far more pleasant idling in the sun 'than being strenuous in some English country house, pretending to be desperately merry, certainly being desperately cold, and spending a lot of money'. At the same time, she was aware of preparations going on in her little hotel for some kind of after-dinner event.

Before she knew it, and after only a little persuasion, she found herself joining a large party of revellers, mostly government officials with their wives and daughters, in dancing to a band on the hotel terrace. She enjoyed the absence of formality – 'one just danced with anyone who asked one, whether one knew them or not' – and learned to tango with one stranger and Charleston with another, before throwing streamers and paper darts in happy abandon. She partied until 6 a.m. and, willing as always to have her mind changed, realised she had rarely enjoyed herself more. Later, on a 'perfect Christmas evening', she swam luxuriously in a secluded bay and drove back 'through the scent of burning aromatic wood, through that queer green, translucent light of unreality'.

When reality did hit, she remembered she still had to work out how to reach her ultimate destination, Portuguese Guinea, a corner of Africa little known to the outside world. For a European colony, it was one of the most difficult places she knew to get to. The best method was to wait for the *Fatala*, an occasional trading schooner run by SCOA,[26] which would take her to Boké, a large French post in the north, from where she could undertake the last week's stretch on foot.

While she waited, she familiarised herself with Conakry, 'a little musical-comedy town where nothing seemed quite real', and yet whose strategic position had made it an object of covetousness by several European nations in the days of the slave trade, 'plied equally by the French and British as by the Portuguese and Americans'. Here too lived many Syrians, some of whom were successful traders in the town at a time when the economic decline of their home country had been exacerbated by the defeat of the Ottoman Empire. She heard about the 'banana problem', which was the main subject of serious conversation in Conakry – the question of how to successfully transport the fruit from West Africa to Europe against competition from the Canaries and elsewhere. She was photographed inside a giant fromager tree, the petiteness of her form emphasised by the vastness of the trunk and, to obtain some relief from the gossip at Conakry where little of outside interest found its way in, she took a noisy train into the mountains of the Fouta Djallon in central Guinea, considered one of West Africa's most beautiful areas.

As she had experienced so often, the boat she was expecting was repeatedly delayed and she had to make do with *La Garonne*, a sturdy, green-painted tub, about 40ft long with a ragged sail, a tiny deck piled high with junk 'and a tiny shelter, six foot square and four foot high, called a cabin'. Despite being accustomed to all kinds of discomfort, as she crawled inside, Dolly was shocked by the state of it. Two sailors crouched among 'an assortment of filth indescribable, of dirty rags and oily tins, articles of verminous clothing and half-eaten, decaying fragments of food. The air was fetid with the smell of vegetable butter and unwashed humanity.' On every surface and ceiling swarmed a host of insects of every description and two large rats scuttled across their feet, to watch them from a corner.

The prospect of spending at best five, possibly ten days here was too much. She gave her newly acquired helper, Suleiman, rags and petrol (considered an effective cleaning product) to clean her quarters, while she went to the chemist for insecticides and returned briefly to her hotel.

Then tragedy struck that would forever haunt her. The news reached her that Suleiman had set fire to himself and the sailors, and all were in the hospital, where she immediately went. She would never forget the sight of one man as he was being treated, writhing in agony while the doctor cut off long strips of ragged, bleeding skin; he died two days later. Dolly found it terrible to think that, 'however indirectly, one has been the cause of a man's death'. It was only she, though, who seemed to take the matter seriously, for 'the lives of black men are cheap in a white man's country'.

Added to her distress was the question of fault and what degree of responsibility would fall to her, with the matter to be settled by the *procureur général*. It transpired that it was not Suleiman who had caused the damage but one *of La Garonne*'s sailors, who had entered the cabin against orders and dropped a lighted match into a pool of petrol. Meanwhile, that night, the boat with all Dolly's luggage on board was stolen by a couple of the crew and she had to borrow clothes from (mostly male) acquaintances.

After four days, she was relieved to learn she was absolved from all blame in the tragic incident and free to go when she chose. Another three days later, with a new helper, Amara, she finally left for Boké on the much-delayed *Fatala*, where she had a cabin to herself and a bed shared with nothing worse than fleas, from where she listened to the chatter in several languages of a party of Syrians who camped outside her window and with whom in daylight hours she shared companionship and oranges.

Up the Rio Nunez they sailed, passing a place where native justice was once meted out. A long pole was still standing in the water, where the prisoner, with arms and legs broken, was fastened at low tide; he was lucky if the alligators found him before the slowly rising river drowned him.

Arriving at last at Boké, she was delighted to find all eleven pieces of her luggage had turned up from *La Garonne* and, although everything was closed as it was a Sunday, the local club had a bar where she could blissfully imbibe a *menthe á l'eau*, long, green and cool, and the British agent of a shipping company and his wife invited her to stay in their comfortable house and fed her delicious meals. In such outpost towns, there were few white women and Dolly had come to admire those who had followed their husbands 'to comparative exile'. It seemed to her that 'the influence of women of the right kind is incalculably to the good – and the wrong kind of woman does not stay long in West Africa!'

For those white women who lived way up in the bush, she had admiration 'that amounts to reverence'. Dolly acknowledged that although she loved the bush, it was 'a vastly different matter spending a few months, however uncomfortably, in a place where everything is new and interesting' and from where, within reasonable limits, she could leave when she wanted, to one spent living day in, day out, 'under strange and often trying conditions, cut off from family, friends and all the cherished habits of a Europeanised lifetime'. A woman's responsibility in Africa was 'enormous' and, she opined, one that British women shouldered particularly well.

SETBACKS

As always, the success of an expedition depended on the planning, so she had little time to spend in Boké, focusing on her week's journey on foot to the frontier of Portuguese Guinea. Despite attempts by the French authorities to dissuade her — great rivers to cross, no boats, mangrove swamps, fever — she left with a hastily constructed hammock and a team of fifteen porters.

The first problem they encountered was a new one for Dolly, a forest fire that presented them with the frightening dilemma of which way to go to escape the asphyxiating smoke. When they had overcome that problem, Dolly was faced with another in the insubordination of her porters, who constantly wanted to stop and rest, drink, smoke and gamble, and demand food they were not entitled to. She thought wryly of the porters in adventure novels — including those she had written herself before she knew better — who were fiercely loyal, always walked in perfect lines at a steady pace and always looked as though they were enjoying themselves. By contrast, when she travelled, 'my *safari* has always something of the air of moving day in an East End slum!'[1]

As they moved deeper into the bush and the country grew rougher, they began to encounter some of the many, mostly Muslim tribes who also embraced an element of paganism. The Foulahs, Landoumas, Yolas and Tendas shared a similar physiognomy and clothing, and their women wore as much jewellery as their husbands could afford, mostly of beads and metal. Dolly noticed curious necklaces made from strings of 5-franc silver pieces that had been unknown in Europe since the war. Tiny bells were another favourite, especially among the children.

While she admired the industry of the small villages, the cruelty of politics was ever present. In one village, Dolly found an election in progress in which the tribal king was an old man who many felt was getting fat and lazy and they wanted to replace him with a younger leader. The measure of how fit he was, physically and mentally, to continue his reign depended on whether he could dance non-stop through a blazing African day: if he broke down before the sun touched the rim of the horizon, he would be dethroned. She watched his vast form shuffle around in the dust, sweat pouring from him, and was amazed later to find him grey-faced but still dancing as the sun set. He had won the election, although Dolly heard later that he allegedly died from a snake bite (a not uncommon death for those who were unpopular) and his cousin took his place.

Dolly's travels in Guinea seemed to present more physical challenges than ever, yet still she was prepared to endure them. Crossing the worst part of the *poto-poto*, or mangrove swamp, had to be done at just the right time, necessitating a four-hour wait for the tide to sink sufficiently low; ahead lay the village of Kannefote, where they could spend the night. They soon found themselves 'knee deep in black, slimy mud that oozed and squelched and sucked one down, while the perspiration poured down one's face and body like tropical rain, and one's heart pounded till it seemed like to break one's ribs'.

Having got out and caught their breath, they realised they would have to forgo that night's sleep because they had to catch high tide on the Compony River so they could launch their *pirogue* or lose half a day's march. As they prepared to bypass Kannefote, they shared with the villagers 'an event of immense domestic importance'. They had been plagued by locusts which were now falling prey to the forest fires that raged on two sides. Dolly watched as their mass turned the sky from blue to dingy grey and realised she had never 'fully appreciated the completeness of that particular plague that affected Egypt in ancient days'. As the singed bodies fell, children ran to collect them, for they made good eating.

Another nightmare greeted them at the river; even to reach the water's edge, they had to cross a wide belt of more mud. 'The next hour was an inferno,' she wrote, 'sweating and straining, balancing on each foot while dragging the other out of the mire and struggling for foothold. We advanced inch by inch.' A faint river breeze carried to them 'a smell of all dead things, of decay and rotting vegetation, that sickened one, making one's head reel'. They grabbed on to whatever seemed solid, sometimes swinging by their arms from the arches of the mangrove roots until Dolly wondered if her arms might break and her back crack in the middle, but 'a slip meant total immersion'. Meanwhile she watched her loads fall into the mud, while myriad insects droned around them, biting savagely.

When they reached the water's edge and launched the *pirogue*, they had to leave the heavier loads behind until they could get assistance from another village. When they arrived on the other side, grounding on yet more mud, some kindly villagers waded out to meet them and two of them seized Dolly by the armpits and carried her to dry land. She was taken to a native house but everything she had, including her bedding, was saturated, and she could not eat because of fatigue.

Dense fog and intense heat accompanied them as they marched the next morning, not helped by Dolly's nausea and aching muscles and the smell of rotting meat, used as bait, coming from the leopard traps along the route. What a relief, then, to be greeted at Diarga by the Almamy Almind'iaye, the chief of a large region, who had worked as an interpreter in Conakry and spoke excellent French. He and Dolly made polite small talk, he cross-legged on a grass mat and she astride his hammock. She found there was 'an immense dignity about these native chiefs in their spotless flowing robes of dark-blue *guinée*, a simple courtliness born of long generations of rulership'.

Food was brought, with huge bowls of rice for her men, fresh eggs, chickens and palm wine. Being rich, Almind'iaye had many wives aged between 60 and 16; most had never seen a white woman and others were not sure if Dolly was male or female and were frightened. One woman, however, who was as fat as Dolly was thin, took a fancy to her, stroking her hands and face and begging her to stay and be her sister. When they left, Almind'iaye insisted on leading them out on his prized possession, a blue bicycle, which had to be carried much of the way because of the swampy ground.

At Sansalé she had a few days of rest when she lodged with a young European trader in a little room off his store, but the forests around were a different story. There lived the *Simons*, who were said to change themselves at will into lions, snakes or leopards, living alone and guarding the roughly carved images of wood that acted as intermediaries with a deity. If someone wanted advice or information, they could take an offering to a *Simon*, and he would commune with the gods for them. However, Dolly was told of people who went to see a *Simon* and never returned and were believed to have been eaten.

It did not stop her wanting to find her own idols, and with a couple of boatmen, she paddled off one afternoon to find a nearby village. Getting out along part of the bank, she had another of the most disconcerting encounters of her life.

Hearing a shout from her boatman, Amadou, she turned to see a huge figure approaching her with a monkey-like gait, wearing only a dirty loincloth. His face was mostly covered with long, matted black hair, his chest was sunken and he had muscular arms and legs. 'His black, feverish eyes glared malignantly from cavernous sockets,' she wrote, 'and he carried a heavy knobbed stick.' She addressed him in French, which seemed to calm him, but then he started whirling the stick near her head. As she ran, her hair fell down, and he realised she was a woman. Catching her, he repeatedly called her 'Holy Virgin' and dropped to the ground, grasping her by the feet and trying to pull her back towards the forest. She talked to him again and she thought she saw tears coming into his bloodshot eyes.

As her men prepared for a quick getaway in the canoe, she wriggled free of his grasp, but now he had 'turned to an incarnation of demoniac rage', with flecks of foam forming in the corners of his mouth. He hurled rocks after her and as they rowed away, he kept pace with them along the bank as they dodged his missiles. As the river grew wider and the rocks fell short, 'A howl, horrible, animal, reached our ears and in the distance we could see a form, scarcely human, that lumbered up and down waving its arms aloft and howling'.

Later, she heard something of his sad story. He was thought to have been a man of culture and breeding but had become 'afflicted with a religious mania, accompanied by a homicidal sideline'. Obsessed with the idea of making human sacrifice to a deity of his own imagination, he had been imprisoned, but he had escaped and for years had been roaming the densest part of the forest and swamps.

Soon, though, she could put the experience behind her because she had arrived at last in Portuguese Guinea. Today, it is known as Guinea-Bissau, after gaining independence in 1974. For four days, she stayed at Cacine, on the shores of the Cacine River, with the hospitable *intendente* and his wife and baby, while word was sent of her arrival to the Portuguese Government at Bolama. It was trying to encourage a new population to the still-undeveloped area by building new huts and charging a hut tax rather than a charge per head as the French did, making it much cheaper to live there, if less healthy. As Dolly observed, 'there is no limit to the number of human creatures who may crowd themselves into a native hut!'

The largest tribe in the region were the Nalous, whose number was in decline due to their custom of not allowing a man to marry a girl of his choice unless he could offer a female relative, preferably a sister, in return. Dolly was intrigued by two boys, aged about 12, who wore skirts of thick raffia, like a ballerina's tutu. They were undergoing the ritual of circumcision and for two years were not allowed to speak to a woman.

She still had Amara with her, having overcome the difficulty of getting him a permit and despite his annoying tendency to borrow money. At Cacine, he came into his own when he removed a family of insects called jiggers from Dolly's foot. They had burrowed their way into her toes and, if not removed, would lay their eggs under her nail; when hatched, the jiggers would attack the foot painfully and take weeks to cure. However, Amara, 'with his great rough hands, handled my sore toes with the precision and delicacy of touch of a Harley Street surgeon as he cut the swollen, hardened skin and removed the offending objects with the point of my clasp-knife'. She also acquired 'a charming godlet, a little black formless female creature, emblem of Iram, the unknown, whom I promptly destined as a mate to the fetish who had bought no luck to the householder of Timbuktu'.

As always, Dolly sought to understand and to convey a sense of the country's history, especially as the Guinea coast was little known to the outside world. In *The Golden Land*, she highlights its ancient explorers and the major incursions in the fifteenth century by the Portuguese, who by the end of the sixteenth century would find their rights in Africa challenged by the British, Spanish, Dutch, French and Danes. The country's complex trading history showed its vulnerability to exploitation and its willingness to trade was recounted by a seventeenth-century English chronicler (although Dolly sometimes doubted his veracity), who found it 'a country where a man may gain an estate by a handful of Beads and a Pocket full of Gold for an old Hat'.[2]

A major part of its story involved slavery, whose evils Dolly describes. While she makes it clear that 'domestic' slavery had existed in Africa for centuries, slave traffic on the Guinea coast 'increased and prospered as the development of America and the Antilles provided an ever-increasing market and we, the British, though eventually among the first of the nations to help eliminate it, were not at that time the least guilty'. Slave-trafficking was lucrative and along the West African coast, with its islands, rivers and lagoons where pirates and rogue traders could easily hide, the practice of slavery lingered illegally.

Despite the country's general lack of development, Dolly did reap the benefit of Portugal's recent road-building programme, which she found to be better than any she had seen in West Africa. It had started in 1917 after what was known, ominously, as a campaign of 'pacification [...] and one by one the tribes were subdued, till now the only spot under military occupation is the Island of Canhabaque in the Bissago Archipelago'.

She may have understated the harsh reality of that event, and she would never know of the brutal war the country would enter decades later to achieve independence. Nevertheless, under no illusions, she summarised its past:

Cruelty, heroism, treachery, cupidity, religious zeal, the soaring spirit of adventure – all these have gone to the making of the Golden Land. A great deal to do yet lies before us, even more than has been done since the days when primitive man of that glittering Coast said of us: 'Gold is the white man's god.'

Soon she was travelling along those new roads in cars sent to take her to Bolama, an island in the Bissago archipelago, to present herself to the government. She was accompanied by the *intendente* of Bolama and an official from Fulacunda, who spoke excellent French and to whom Dolly was grateful for his kindness, patience and humour on their journey. As they motored north-east from Cacine to the Rio Grande, her record of the tribes they met in the villages, some of whom she had first encountered elsewhere in Africa, further exemplifies her interest in and knowledge of ethnography.

Soon they were among the Foulah-Fiorros, 'the best class of local Foulahs [...] a pastoral people, hard-working and generally rich'. While the women wore plaited hairstyles 'curiously like our modern shingle, intertwined with bright red beads and small shells', she was unhappy with the way they were 'deplorably overdressed'. Their upper bodies were covered 'with loathsome bits of European clothing – by order of the Government, I was told, who are trying to instil a sense of the unseemliness of partial nakedness into a people who hitherto have possessed no false modesty'. Almost everywhere in the country, 'where white authorities were in contact with the native, my poor little sense of aesthetic values was offended by this same governmental decree'. Although many 'good Colonials' had told her she was wrong, she could not help but think that whereas a 'primitive black woman unadorned is often a fit subject for Praxiteles [an ancient sculptor of the female nude], partially clothed in a [...] camisole of scratchy cotton and coarse lace, she is a broken melody, a pathetic figure of fun'.

Some tribes she already knew, such as the Mandingoes, for whom she had a 'sneaking affection' and who, in Portuguese Guinea, had lost land to the Foulahs. They still lived under a system of totem or clan, each identified by a symbol of a plant or animal.

Animal life was not something Dolly often commented on but in Portuguese Guinea there seemed to be an abundant variety: hippo in the rivers and even in the salt-water channels of the islands; different types of antelope; monkeys, ranging from the small and chattering to large, dog-faced baboons that barked like dogs and were 'powerful brutes, sometimes nearly the weight of a man' and nasty to meet alone if they were in a bad temper. The birds were 'things of radiant beauty. They pass one like streaks of lightning across the dusty road.' One resembled 'a small red flame', another 'glinted in the sunlight like burnished steel'.

She saw 'little green parrots and great pompous marabout storks, and aigrettes, whose precious feathers one buys by the weight, at eight escudos the gramme'.

Of the insect life she had less to say, other than mentioning the two worst types of mosquito, one that caused malaria and the other yellow fever (for which there would be no vaccination until the 1930s). Fortunately, in Portuguese Guinea she escaped malaria, and indeed any serious illness, and found the temperature generally less oppressive than other parts of West Africa.

They reached Bolama, the most important island in the group with its 'pretty, sleepy little town', after a short crossing from tiny São João. She was to meet the governor but first settled into the only hotel, which was of a type she had stayed in before, where the large, lofty rooms of the former private house were separated only by a canvas screen. She had already ascertained that only men went outside, 'for rarely did one see a white woman of any class in the streets of Bolama or any other Portuguese colonial town'. However, even though Amara would be with her, she had to make it clear to those who tried to persuade her otherwise that she could not be expected to remain in the hotel all day, for her curiosity and restlessness 'demanded gratification' and her 'British liver, exercise'.

Her point was graciously, if reluctantly, conceded and she was relieved that local politeness meant she was not stared at, even though she remained acutely aware that Portuguese 'ladies of quality' never walked out unattended. At the same time, she found much helpfulness and patience when it came to the language problem, more so than in many countries. She much regretted her inability to speak or understand Portuguese in this country that was entirely new to her, especially as it proved an obstacle to her communicating directly with her own sex. While most Portuguese men she met spoke French, she only met one woman who did.

The Governor of Portuguese Guinea, Major António Leite de Magalhães, was charming and courteous and gave 'whole-hearted assistance' to her projects. Nevertheless, 'there were quite a lot of things I wanted to do that others did not want me to do', although she averred that in everything reasonable or possible her requests were acceded to, however troublesome. While there must have been an element of her discovering what they wanted her to discover – much information came from members of the Cabinet – in the bush, she was sometimes able to leave her host and wander off alone.

The native people she met, not only in villages but 'in the wildest of the bush places', seemed cheerful, 'with no fear or avoidance of white people'. However, the appearance of happiness may have been deceptive. While she was told that certain peoples, including the Manjacos (then the third-biggest tribe), were 'taking easily to the ways of civilisation', the veracity of such statements doubtless depended on their source. The Manjacos had enjoyed autonomy for centuries until 1913, when a military group comprising Fula and Mandiga soldiers led by Portuguese officers attacked and conquered their territory.

One day, she came across a group of young men of the Balanta tribe, mostly farming people and the largest ethnic group in the country, and she was delighted to find them in their habitual state of near-nakedness, without the European clothing she so despised. When they saw her, a white person, and holding a camera too, they grabbed the clothing they were expected to wear, but they soon understood from Dolly's gesturing that there was no need and posed naturally. 'So, for almost the first time,' she wrote triumphantly, 'I got a photograph of my bush-brothers looking like themselves and not the leavings of a junk shop!'

There was no class distinction among the Balantas, all being 'free and equal'. They believed God to be far away and reachable through spirits and sacrifice. For them, death was not a natural phenomenon but the act of an 'eater of souls', and while the deaths of old people were accepted more placidly, that of a young man would cause his family and friends to spend days trying to wake him until bodily disintegration began, when he would be buried at his house and other rituals observed.

The Balanta woman kept control over her property on marriage, while a divorce had to be the subject of a hearing before the *intendente* and a council of village elders. Dolly sat in on such a hearing where, through an interpreter, a 20-year-old woman explained that in her four-year marriage, her husband had beaten her and then thrown her out with their baby. She went to live with her father, then found another man who loved her. Now her husband wanted her back but would not take the baby. Witnesses were brought in, the husband had his say and, after finances were discussed, she was granted her divorce. The *intendente* told Dolly that the two chief causes of misdemeanour and ill-feeling were "'cows and women … and the cows are the more important'".

Dolly drove further north with her host, the monotonous flat bush giving way to forest then open savannah, passing a medley of tribes on the way. A riverboat took them through more mangrove swamps to São Domingos in the north-west and on to the Casamance, 'the strip of Senegal that lies between the British Gambia and Guinea', which formerly belonged to Portugal but now was French. At Ziguinchor, a major port in Senegal, there were traces of the old occupation, where Portuguese Creole was still understood, although to Dolly's relief here she was able to converse in French and realised what an 'unmitigated bore' she was in her ignorance of Portuguese.

Back at São Domingos, they journeyed deeper into the forest, driving for miles without seeing a human or any sign of habitation, until unexpectedly, they would catch a glimpse of one of the most remote tribes, the Felupes, who were aboriginal to the country and 'the least under the domination of Portuguese civilisation'. Dolly was privileged to see them and was surely one of very few white people to do so. She wrote:

> Sometimes from behind the bushes would appear slim, muscular men, young for the most part, naked except for a scanty loin-cloth, with many tattoo marks and a quantity of rough metal bracelets, their hair plastered down with black mud, encrusted with small, round pieces of metal. Each man carried a big bow and arrow, and sometimes, when they had lost their preliminary timidity, they ran after our car, leaping into the air, and shouting and laughing.

Although there was some Islamic influence upon their beliefs, they were mostly animistic, believing in a universal but remote god, whose intermediaries were the fetishes kept in their houses and invoked by the sorcerer in time of need which, as they were agriculturalists, included protection of their crops. To these fetishes they offered bowls of palm wine and the blood of goats or pigs, which replaced their ancient practice of using the blood of their enemies killed in war. They believed in a future life, with good people going to a paradise, the bad passing into the bodies of animals such as hyenas, crocodiles and snakes; if one of those animals did any harm, they thought it must be inhabited by the spirit of a dead person, for they did not believe that animals of their own accord were dangerous or cruel.

In the forest of the Felupes was Susannah, which Dolly considered 'one of the loveliest spots I have seen in West Africa, where there are many lovely spots'. A few government agents lived in a small compound where Dolly was to sleep, amid forest that was 'as untouched as it was two thousand years ago'. She wandered alone around the nearby village of sturdy huts and found, unusually, that the women took fright and hid from her, confirming her suspicion that she was the only white woman they had seen. When curiosity overcame their fear, Dolly was surrounded by dozens of them, their heads close-shaved, dressed only in a belt of beads and leather passed between the legs and a quantity of jewellery made of rough beads. It was little surprise that she was an object of intense interest, with her skin and hair inspected and her skirt lifted. Everywhere she went in the village they followed, and every time she tried to leave they tugged at her and invented games and tricks to grab her attention, so that she had to resort to using her camera as a defence, as the sight of it made them retreat.

On she went with her host (and Amara, still with her) for a tour of the Bissagos Islands, the archipelago about 30 miles off the coast of Portuguese Guinea, a beautiful and wild area with a tragic past. Its people had formerly been 'carried off in many thousands to the Antilles and the United States, and it is said that to avoid slavery, four-fifths of them committed suicide on the journey', chiefly by drowning. Unusually for black Africa, among the Bissagos women, 'feminism reigns supreme'. It was the woman who selected her mate and courted him and decided when she was materially ready for marriage, and when she was tired of him, she simply threw his possessions out on the street during his absence, which was the sign that he should consider himself divorced.

In a house in Bubaque, Dolly spotted 'a fetish after my own heart' tied to the foot of a bed, a formless object of dark wood, with four legs, no body and a bird-like head with a long crest. Thinking it would make a worthy mate for her little lady fetish from the Soudan, she asked the owner if she could buy it, but he refused, as it was an intermediary between himself and Iram and he feared bad luck would befall him. However, the next day he turned up with an interpreter and explained he had changed his mind, because he had taken a fancy to a sparkling hat brooch Dolly wore: he felt certain that a glittering object from across the seas 'would assuredly be more acceptable to the great Unknown than a mere wooden idol of a kind to which he was accustomed'. So it was that a little present Arthur had bought her 'for a few francs in the Rue de Rivoli' found its way to a dark hut in the West African jungle.

Dolly was aware that white people were not welcome everywhere in these islands. In Bubaque, she was reminded of the power of the African tom-tom, which reverberated around the village as she wandered, with its 'language as definite as spoken codes' and yet it could only be understood by the initiated; she had already discovered elsewhere in Africa that often the next village knew in advance that she was coming, thanks to 'the wireless of the bush'. Now her guide told her, on being pressed, that the message concerned a secret meeting of the elders, but Amara told her later that the men said the drum also spoke of her, because she had gone into their medicine house and a sacrifice would be made in atonement.

When she wanted to visit the island of Canhabaque, permission was given only reluctantly 'on account of a possibility of trouble', because the islanders had refused to pay the hut tax and become rebellious, resulting in a small company of native troops under a white lieutenant being stationed on the coast. While there, she inadvertently frightened an islander, who 'fled as if all the devils of the underworld were after him', and then, as she went to help up a little boy who had fallen and was crying, his mother, 'eyes wide with mingled fear and fury [...] came for me like a wild thing', grabbed her child and ran.

Later, an entire village 'evacuated itself at our approach as quickly as water runs from the bath' until rumours of presents of tobacco trickled through and they gradually came out, the old women first, as Dolly found they always did in the African wilds; they were the most devoid of fear, 'secure, I suppose, poor souls, in the knowledge that they are neither rich, nor young, nor beautiful, and therefore safe from any species of attack!'

Dolly understood the people's fear, for there had been a bloody battle with the Portuguese on Canhabaque just a decade or so earlier (the island would not officially come under occupation until 1936). Beside the root of a giant baobab which the islanders used as cover 'when they lay in ambush for their attackers', she could still see the marks in the bark where they had rested their newly acquired rifles, delivered to them 'by contraband' by white men. But the paradisiacal beauty of the island belied such events and as they left in a sunset of amber and rose, Dolly wished she could have stayed longer in 'those strange, sleepy islands, enervating and beautiful, that caught one up in their queer, savage charms like the breath of an opium dream'.

Preparing to return home this time involved a good deal of crating up her newly acquired masks, carved animal heads and other idols. As she did so, not only did she muse on how they would look on her blank wall in her London flat but on a belief in magic in Africa generally. In *The Golden Land*, her chapter called 'The Dark Gods Still Linger' considers the fusion of magic and religion, 'born at the same time, of the same material, in the hearts and minds of the same people, possessing the same needs', and how science fitted in.

In discussing the importance and use of different items, she points out that the mistake many travellers make regarding fetish idols is believing that someone worships or reveres them as gods. Rather, they were worshipped 'as intermediaries between themselves and the vast, unknown, supreme god'. The object gives the person 'something tangible they can talk to, even though only symbols. The spirits of their great men too, they revere as intermediaries whose mediation may befriend them.' She draws a comparison with Catholicism, where devotees ask figures of saints for intercession and burn candles, and she takes the reader back further to pagan times, where various forms of mediator carried up prayers to the gods.

Dolly never dismissed the power of belief, or indeed the possibility of what some might call 'magic', because she had experienced apparently inexplicable events herself – a healthy young man she knew lay down and died within a week because a witch doctor said he would, and this had a deep impact on her. Of idols she wrote:

Personally, modern sceptic though I may be, and hardened too with much travelling I cannot quite regard the African fetishes as just senseless things of wood fashioned by ignorant savages. I think that the strange unknown gods of nature, gods fashioned by an eternity of primitive belief, have breathed into them something of their own mysterious essence.

Dolly's return to Dakar took her via the Gambia for a fleeting visit, where she witnessed her first shooting of a lion. In those days when trophy hunting was still big business, and while she was of her time in tolerating it, she acknowledged that the reason her host's Gambian servant refused to take part was because the totem of his clan was a lion: if he lifted a finger in any way, he would have to make offerings of kola and do a dance of atonement to his totem.

Her departure from Dakar was unexpectedly ceremonial, with the townsmen and their wives lining the wharf alongside a group of fifty tall Mandingoes dressed in ceremonial costumes and a quartet of men drumming, while limousines drew up and disgorged uniformed men and pretty women in smart frocks. However, the fuss was not for her because sharing her ship was the Governor-General of French West Africa and his consort.

As they pulled away and she looked fondly at the mass of humanity on the quay, a 'queer mixture of north and south, of black and white', someone asked if she would like a drink. 'I did not want to see Africa once more die behind me,' she wrote poignantly. 'I did not want to feel that catch at the back of the throat, to feel that feeling of emptiness [...] I went in for the apéritif.'

While Dolly was away, British journalist Lady Drummond-Hay wrote a feature on American women. While she acknowledged their strengths and the material and social advantages they enjoyed compared to their European sisters, she wondered where, in contemporary history, were the American versions of Lady Dorothy Mills or Rosita Forbes? True, they had aviator Amelia Earhart, whose name would go down in history, 'but Europe and the east can boast names who have made and are making history'.[3]

On Dolly's return in mid-April 1929, the press mainly wanted to know about magic and witch doctors, who she said had real knowledge of medicine and were some of the finest herbalists in the world. Her manuscript of *The Golden Land*, which she had been writing during breaks in her journey, went off to the publishers and other features appeared, including one that busted the romantic legend of the *sheikh* and another in which she confessed to vanity in giving herself a home-made face pack in the West African bush, before giving up her only jar of skin cream for use on a baby.

Her reputation was at an all-time high, even in unexpected quarters. She was mentioned in an address to a large Scout and Guide parade on St George's Day, attended by Lady Baden-Powell, the association's Chief Guide and wife of the founder. 'You girls may have dreams that carry you far beyond the factory and the shop,' the vicar conducting the service said to the Guides, which alone would have made Dolly happy:

> You do not stop at hoping to be a telephone operator or a shorthand typist; your particular heroine may be a Florence Nightingale [...] and you would like, no less than a boy – for girls can do almost everything the boys do nowadays – to be an explorer like Lady Dorothy Mills, who practically alone has penetrated into the heart of Africa to explore its secrets.[4]

That year she was the only woman among eighteen explorers and other risk-takers to feature in the second book of a four-volume compendium, *More Heroes of Modern Adventure*.[5] The chapter about her, written by one of its two authors, referred to her explorations generally but, since it was 'out of the question even to outline all Lady Dorothy's adventures', focused on those in Liberia. 'Always she goes alone,' it said, 'and she has that curious, almost uncanny knack of handling savage and uncivilised men which is the hallmark of the true explorer, and which has brought her through in safety out of all kinds of tight places. And she can always laugh, even at herself.'

That ability would be gravely tested two months after her return. In June, she went to Royal Ascot as usual. On the way home, when her taxi was just minutes away from Ebury Street, it crashed. All she remembered was 'a great car rushing down [...] two hours later, of lying alone on my bed at home, of a lacerating pain in my head, and a pool of blood'. She had sustained a fracture to the base of her skull and facial injuries: initially, it was feared she may not live. Fortunately, her condition stabilised quite soon, and she, or someone on her behalf, was able to issue a statement via *The Times* on 3 July, thanking all the friends who had made enquiries or sent flowers, and assuring them she was making good progress and hoped to be convalescing the following week.

But while the tone was upbeat, she was not. Months of physical and mental pain followed, exacerbated by the cruel irony that, having survived in so many wild places abroad, she should be injured on her doorstep. Her world had shrunk to 'a darkened sick room, to ice-packs, nausea, a scarred face, to intolerable pain and a humiliating weakness'.[6] Just months earlier, she had said her chief dread of old age was that it would involve her sitting in an armchair and merely reading the accounts of other people's travels: she must have thought that time had arrived prematurely.

ALONG THE ORINOCO

News of his daughter's accident reached the earl in New Zealand soon after the event, although not from family members. 'I had an alarming telegram from a friend in London,' he told Dora at Wolterton, 'that Dolly had been terribly smashed up in a motor smash. I was glad to hear this morning that she is going on satisfactorily.'[1]

In her autobiography, completed about fifteen months after the accident, she writes with honesty and dark humour about her state of mind and what her experience taught her. She contemplated suicide with 'a detached morbidity', but realised she could not financially afford to die in the manner she would have liked, 'So I continued to live, and now I am intensely glad!'[2]

One thing she learned was fear, which, for some reason she could not fathom, she had never felt until she was hurt: it made her realise she had been unsympathetic towards other people's fears. Crowds now frightened her, as did anything rushing past. For months, driving in any kind of vehicle was 'a long drawn-out torture that was intensified by my frantic efforts to conceal it'. In a busy thoroughfare, she had clung on to a lamp post while buses swooped by, until helped by a policeman.

Now she was glad she knew fear, 'for it has taught me there are greater things between me and death than my own self-assurance, that I am lucky to have been given the sporting odds on being alive at all'. It gave her greater sympathy for the experiences of others, however dissimilar, 'For each heart knows its own particular form of hell, and each has to be borne alone'. She had learned gratitude too, which she hoped would make her 'a pleasanter and more useful member of society, less of a hedonist, to make better and more appreciative uses of my assets'.

Before reaching that stage of self-knowledge, however, she had 'a long and infinitely trying convalescence [...] periods of pain, constant setbacks and discouragements [...] the humiliating consciousness of inferiority, the practical and mental worries resultant upon one's disabilities, and the knowledge that one is worrying others'. In November, the earl updated Dora, 'I hear from Dolly that she is getting better from her accident. She has had a lot of misfortune.'[3] However, later that month she had to publicly announce that doctors had forbidden her to fulfil any engagements and among those she had to cancel was a talk in London in aid of a hospital charity on her experience of magic in Africa.

Arthur's role in helping her recover must be presumed. He was still writing copiously. A month after the accident, his feature was published about his earlier visit to Devil's Island, the feared penal establishment of French Guiana. There, he had met an ex-convict who described life as suffered by another prisoner, Alfred Dreyfus, the recent victim of one of the great miscarriages of justice.

Arthur was also turning out short stories, reporting on polo matches and had recently completed a comedy and a romance, which was serialised in the press. *Intrigue Island* was billed as 'a story of love and intrigue in the glamorous east' and led to some flattering coverage. 'Professional soldier, writer and world-wide traveller, and one who has faced danger and death [...] in the remotest corners of the earth, Captain Arthur Mills provides one of the most romantic figures of the present day,' said the *Gloucester Citizen*.[4] Dolly was not overlooked:

> None the less romantic is his wife, Lady Dorothy Mills, with whom he shares a friendly rivalry in the most exciting of adventures and the output of realistic fiction, and who is one of the few women who has sacrificed home comforts for the lure of the jungles and desert wastes of tropical lands.

Those wild places were unattainable at that moment, although the accident had made her 'a bolder gambler', having learnt that in life, as in other games, 'the element of chance is almost more important than that of skill'.

If she was able to be cheered at all in those miserable months, then the publication in November of *The Golden Land* and its reviews would have helped. As pointed out by the *Illustrated London News*, 'All over the world now the question of the relations between the white and the coloured races is becoming more and more insistent', and there followed examples of such countries. 'This question of colour, and the general study of native life in remote lands, often forms the main interest in books of travel.'

At the time of the review, January 1930, the voyage of the Prince of Wales (the future Edward VIII) to Africa to hunt big game had turned the reviewer's attention there (although not on the subject of hunting), and he selected three books on aspects of Africa, one of which was Dolly's. This 'well-known woman traveller [...] writes with her accustomed charm and vivacity,' said the review, 'giving fresh evidence of her insatiable zest for wandering [...] She herself evidently has the faculty of making friends with the natives. She gives many vivid pictures of their ways of life, as well as memorable glimpses into their history.'[5]

The Scotsman found it strikingly illustrated and 'entertainingly written, and instructive in its observations on the foreign methods of colonisation [...] attractive reading to travellers contemplating a similar trip and to readers generally who relish vivaciously told stories of enterprising travel in the West African wilds'.[6] The *Daily Mirror* said, '[She] writes with all the verve and freedom of good conversation that is really informative and draws on considerable stores of knowledge.'[7]

In the spring of 1930, she went to familiar Tunisia to recuperate and came to feel that she was really meant to live, as though she had a 'guardian angel' who was taking an active interest in her fate. In Tunis, she had an encounter with a man called Messaoud, whom she knew slightly and who, to her great embarrassment, declared his love for her. When she made it clear she was not interested, his insistence became threatening, although Dolly thought he had got the message. Later, she was approached by a small boy who gave her a delicious-looking melon as a present, telling her it was good to eat. She put it aside and a few days later went to use it in cocktails for local friends. As she went to add the fruit, her friend Abderrahman exclaimed in horror that it was wild melon, 'the rankest poison', and would likely have proved fatal. Subsequent enquiries revealed that it was Messaoud who had instructed the boy to find her and persuade her to eat it; unsurprisingly, the man had not been seen since.

That was not the only time she felt she was saved. Her injuries caused her occasional dizziness, so she had taken to using a walking stick for the first time. Returning to her hotel a few days later, in a quiet street in broad daylight, she was approached by a man who made as if to hit her over the head with one hand and steal her bag with the other. She managed to jab him hard in the stomach with her stick and as he doubled over in pain, she dived into an archway to hide. When she emerged, he had gone. A friend told her she was lucky to escape without being stunned and her bag stolen, as foreign women were a particular target. This time, she reflected, 'my guardian angel used the very means of my weakness for my own preservation'. Again, she was glad to be alive, to have the chance to do more things and to do them better. In essence, 'to pick up the threads of a life that nearly left me [...] to build up, like the ant, the fragments of a wasted year'.

By July, her growing strength and improved mental state saw the return of her wanderlust, although doctors said she should not contemplate travelling yet. She pored over maps and threw titbits of her thoughts to curious journalists who were keen to be the first to hear her plans. Tibet was a possibility, especially mysterious Lhasa, to travel in the land of the Mongols.

By now, she was gradually socialising again and attended a relative's wedding, where, as they danced, she became disturbed by a strange aromatic smell and began to feel faint, although she had never fainted before. She was taken home and the next day felt fine. Shortly afterwards, she received a letter from Abderrahman saying Messaoud was believed locally to have 'gone a little mad' and to have put a spell on her. Although he knew she was sceptical of such matters, Abderrahman advised her to take care, especially at the time of the full moon: when she checked, she found there had been a full moon on the date of the wedding.

That she was the victim of some kind of malevolent influence seemed to be confirmed by a man who had seen references to her putative Tibetan journey in the press and introduced himself as a Tibetan and a llama. Invited to Ebury Street (where it was impossible to avoid seeing her collection of masks and idols), he began talking about an evil that surrounded her and how the soul of another was fighting for hers. Recounting this in an American newspaper, Dolly said that while she had seen things in Africa that could not be explained, she was strong-minded and unafraid and believed that 'only by my own will can the spell be exorcised. I do not listen to the dark voices.'[8] A scientist invited by the newspaper to contribute his opinion on her experiences wrote about the power of suggestion and the wiles of the so-called magician or psychic in picking up and using information about the victim from clues dropped unconsciously.

While the press awaited her decision on her destination, they could satisfy themselves with previews of her autobiography, *A Different Drummer*, due to be published in early 1931. Although it is a fascinating read, it avoids the intimate and instead focuses largely on aspects of her explorations beyond those described in her books, spinning encounters with interesting individuals into what are effectively short stories.

Editors had a wide choice of subject and mood to whet the appetite of their readers. For the human and humorous aspects, one might dip into her chapter called 'Love Adventures in the Wilds', which included some marriage proposals she had received: the Arab who wished to marry her because her eyes were light; the young Corsican who 'poured his passion into the largest trombone ever seen'; and the Frenchman in a rickety canoe on a West African river who, as they were sinking and trying to bale out, irritated her by repeatedly proclaiming his love and telling her she was beautiful, 'for at no time am I beautiful, and at the moment was looking quite my worst'. Touchingly, she received a proposal by letter from a young Swedish farmer she had never met, who had read one of her books. It appealed to the call of the wild in him, making him feel they were kindred spirits, and he urged that they should 'pursue its siren voice together'.

For those curious to know how men behaved towards a lone woman in wild country far from home, her chapter 'Chivalry in the Wilds' took an insightful look at men and their conduct. In some, the chivalric instinct was 'as fundamental as a sense of personal honour; in others of equal education and apparent breeding, it is but a veneer of training that cracks under a hard test'. Tellingly, she observes that it is not possible to tell a man's 'real attitude towards femininity till all the restraining and modifying influences of civilisation have gone by the board and one is up against the stark realities of life and death, hunger, danger, passion and ambition. And then one has some great surprises.' She had met some 'bad and violent characters' from whom she had received kindness and 'real acts of chivalry' – those who she could not ask to tea in her London drawing room, 'but with whom I would unhesitatingly take a trip into the blue'.

In most wild countries, 'where feminine independence is not yet admitted', the solitary female travelling was 'apt to be misjudged by civilised men', and if she had any degree of personal charm she was certain to receive 'gratuitous insults and sometimes to find herself in equivocal situations that are the best unpleasant and at the worst extremely inconvenient or even risky'. By contrast, 'The rough man seems to make few such mistakes. He gives respect where respect is due, not of necessity but from instinct' and he will treat the woman he respects as a priceless piece of art. One of the most chivalrous men she ever met was a Spanish lorry driver.

Dolly experienced her share of horrors. In Tangiers, she was invited to a house by some male visitors to her hotel who she had believed to be friends. When she smelled the reek of liquor and heard rough voices in a language she only partially understood, followed by groping hands, she realised she was in a nightmare scenario and ran 'panting and sickened, through strange tortuous streets, back to the protection of lights and crowds, back to the shelter of my hotel'.

She found that it was 'in the Latin dependencies that a woman most comes up against the sex problem, and where one's Anglo-Saxon lack of interest in it is least understood'. On the borders of Tripolitania (Tripoli), an old French colonel with whom she had struck up a friendship asked her if she had yet deceived her husband. She told him she had not, which he found hard to believe, as he said she must surely have had lots of opportunities while travelling alone. Laughingly, she told him she had had so many opportunities that it left her without any inclination for it, which he found bizarre.

Sometimes the 'sex element [was] a very real nuisance, in countries mostly under military administration', where one's safety, means of transport, even means of livelihood were 'so absolutely under the control of men who sometimes do not hesitate to use their power in their own interests'. Dolly recounts the chilling story of being in North Africa and needing camels and a specific number of men so she could continue. The commandant, who she had 'the ill-luck to please', would only give them to her on his own terms. When he was suddenly called away on official business, she showed the love letter he had written to her on official paper in French to the local *caid*, taking the chance that he would not be able to read the language. Impressed by the seal and believing Dolly's explanation that it was the commandant's instruction to give her the resources she needed, he obliged. She was three days away before he returned and discovered the hoax.

In her chapter 'Love in Africa', she shares romantic stories of diverse women she met, from the joyful and amusing to the tragic. Among women of all colours, 'the same problems as have puzzled white women throughout the generations' were still relevant.

In November, she had the honour of being elected a Fellow of the Royal Geographical Society, which was particularly gratifying because women had only been admitted as Fellows since 1913, following a twenty-year debate. Her Certificate of Election stated, 'She has travelled extensively in both West and North Africa off the beaten track and has written several travel books. Her many articles in the best newspapers and magazines are well known.'[9]

At around that time, she and several other well-known figures were asked what they considered to be the secret of happiness. For Dolly, the answer lay in one word: 'Freedom: not only freedom for action, which is largely a matter of chance and opportunity, but freedom of thought.' By that, she meant freedom to 'develop one's personality, to travel if possible, and to broaden one's outlook by learning how the other man lives'. She was happiest when travelling, and when she was worried and depressed, she said it was Africa she thought of. Although she knew her preference for travelling alone seemed selfish, when she returned she had a 'much fuller appreciation' of her friends, and she was readier to appreciate their qualities and overlook 'their foibles and little weaknesses'. Mental or physical restriction was bad for people, making them restless and unhappy, which was then transmitted to those they encountered. To married people, she said, 'Live your own lives and make your separate friends.'[10]

Nevertheless, she and Arthur did socialise together sometimes. In December, they attended a glamorous party, whose guests gave a glimpse of pre-war society. Alongside earls and ambassadors were some of the former Bright Young People, including Mrs Bryan Guinness, born Diana Mitford, one of the famous sisters. She had recently married the brewing heir, but in two years' time would become Mrs Oswald Mosley, wife of the fascist leader. It was the last party Dolly would attend for a while.

On 18 December, as Arthur's latest novel *Escapade* was serialised in the *Daily Mirror*, Dolly left Plymouth for Venezuela for four months. For centuries, gold had been its irresistible, often deadly attraction, but in 1914 a new source of wealth had been discovered, making her want 'to linger where oil – the world's great motive power – bubbles from the sun-baked earth, to share the thrill of the wide llanos, land of strong men and swift horses, to learn the mystery of the great waterways of the Orinoco'.[11] Little had been written in English about that corner of South America: it was the first-discovered part of the new continent, yet, at that time, still the least known, 'a land of old adventure and romance, of piracy and dreams, and deeds of derring-do'.

On arriving at the port of La Guaira, she faced more bureaucracy than ever, completing numerous forms declaring, among other things, that she was not suffering from 'leprosy, trachoma, insanity'; had not committed a crime or offence punishable under Venezuelan law; would not propagate 'the assassination of foreign public functionaries [... or] disturb public order [...] or compromise the Republic's International relations', and so on, all before she and her fellow passengers had been visited by the doctor and emigration officer 'and all the other visits of a South American landing'.

Eventually, she was taken under the wing of official friends to whom she had been recommended and they took her to the capital Caracas, a feat of engineering 4,000ft up in the mountains along roads built under General Gomez, de facto ruler of the country, who she would meet. Combating what she realised was altitude sickness, she managed to appreciate the extraordinary views down into the lush green valleys and out to the coastline, with cool and spacious Spanish houses standing among green fields and tall patches of sugar cane. In the warren of little streets, she found a diversity of people, some with traditional Spanish features, some with a mingling of white and Indian, with high cheekbones and sleek black hair, alongside Syrians, French, Germans, Jews, Italians. Here, in his birthplace, she saw the first of a multitude of statues to Simón Bolívar, the eighteenth-century liberator of Venezuela (and Bolivia) from the Spanish and something of a folk hero.

At Maracay, Dolly was one of a small group introduced in his hacienda to General Gomez, who was accompanied by half a dozen army officers, while 'lean brown men in the greenish grey uniforms of the Venezuelan Army walked about, martially'. Gomez spoke briefly to her in Spanish and seemed impressed that she was a writer.

A 74-year-old dictator, he had 'quashed the revolutions that decimated the country [...] When anyone grumbles, they grumble under their breath,' she wrote. He was acknowledged to have done much for the stability and infrastructure of the country and was 'open-minded to foreign enterprises', yet Dolly felt that 'behind the keen eyes one sensed power, acumen, reserve, maybe a hint of cruelty', with which his enemies would have agreed. He was not married and was 'feudal in his ideas and habits [... and] believes in the *droit de Seigneur*' – certainly, he had had two mistresses and dozens of children by them and others.

Few Europeans had been to Maracay and Dolly's description would be of interest. The town's magnificent Hotel Jardin was a concept of the general's and was government-owned, a place where the rich and fashionable came to drink or stay and catch a glimpse of the great man. 'About everything in Maracay there can be dimly felt an air of constriction and formality,' she recorded:

[It] is the local Potsdam, a Potsdam of court and army created by the General, where matters of state are settled, where heaven knows how many revolutions are nipped in the bud, where men speak warily, yet nevertheless seem to know a great deal about each other's concerns.

While romance may have spurred her desire to see a country – its natural beauty, peoples, millennia of myths and stories – her view was never rosy. At Puerto Cabello, on the coast that was part of the old 'Spanish Main', she heard stories of the prison, which seemed 'to stare at one with malignancy as if the fierce soul of it regretted the old days of man's inhumanity'. The coast was one of 'old romance, of feats of heroism, of dark deeds, of violence, of strange cruelties; of romance too, of dreams and towering ambition'. Buccaneers, who were French, Dutch and British had been well known here, all united in their hatred of Spain, especially the British. The best known were Raleigh, Drake and Hawkins, 'men of violence,' said Dolly, and yet whose actions were distinguishable from the gratuitous barbarism of 'monsters' like Henry Morgan or the pirate François l'Olonais. Neither did Spain cover itself in glory in those early days of colonisation, having a tendency 'to ruthlessly exploit and tyrannise'.

As always, her eyes soon fell longingly on the horizon. Despite the usual discouragements about transport, low water, non-existent man-power and the like, she wanted to go south to the Orinoco River, known as the 'black river', where Christopher Columbus realised he had dis-covered another continent. This time, though, was added the frequent and ominous warning, 'Go to the Rio Negro and die', referring to its fever-stricken upper reaches.

One thing she had learned from all her travelling was patience. She knew that sooner or later, she would see 'the great lost world of forest and water, and particularly my brown brother who inhabits them, with his beliefs and habits so different from the black men of other jungles I had travelled'. Eventually she *would* follow the trail of El Dorado, the Golden City, sought in vain by Sir Walter Raleigh, 'for the most beau-tiful of all are the cities that have never existed'.

She decided to go north to Maracaibo, a principal port and the centre of the quickly developing oil industry and was pleased that the only form of transport available was a motor bus: experience had shown her that it was 'the encyclopaedia of the people's soul, its promiscuity is richer in information than a dictionary'. It helped her Spanish, too, for the little she knew had been learned in polite society in London and was not sufficiently vernacular for use in Venezuela. Some less-welcome situations were familiar in the towns, particularly hostility from land-ladies who were suspicious of women travelling alone and unannounced, considering them to be of doubtful virtue. They stopped for the night in Valencia, the second biggest city, where she was initially refused accommodation: the telegram booking her room had not arrived and the dour-faced landlady told her (in French) that there were no vacancies, even though it was clear she had no other guests. Dolly eventually persuaded her that she would not bring her establishment into disrepute and paid the price of being permitted to stay by being woken up half an hour after she fell asleep by a torch shining in her eyes and a demand for payment in advance. She was fortunate that it was the only truly inhospitable experience she had.

With two days to wait at Valera, a trading centre near the Andes foothills, she wandered its hot, rambling streets with her camera and practised her Spanish on friendly strangers. She was struck by the fine looks of the people 'with their Spanish-Indian blood', and noticed a younger man who she thought would have been a great hit in London or Paris, because he looked just like the late Rudolf Valentino or Ramon Navarro, 'with the same pantheresque figure and poise, the same black wells for eyes [...] the same conquering smile and general air of a romantic feline'.

For the final part of her journey to Maracaibo she was invited to sail along Maracaibo Lake on a luxurious yacht belonging to the Caribbean Oil Company, where despite the trappings of civilisation – the delicious drinks, the friendly faces – she was soon reminded of where they were. Not only had the heat changed from dry to damp, steamy and devitalising but she was aware that in the bush behind the opposite bank lived a tribe of Indians whose faces few people ever saw.

The Motilones were one of the last of the primitive tribes, whose weaponry consisted of long poisoned arrows. As there was oil on their land, 'more than one employee of the oil fields has been killed,' said Dolly; apparently, two were shot not long before her visit, and although one was saved, the other died from a haemorrhage soon after the arrows were withdrawn.

At the last minute, she had taken on a cook and general help called George, 'a lady-like' young man, who was a British subject from Trinidad, a non-drinker and non-smoker, 'with a keen desire to improve his mind, and whose chief recreation was getting me to tell him Bible stories'. Eventually, they reached the Orinoco at Ciudad Bolívar, where they boarded a boat called *El 3 de Diciembre*, manned by a motley crew. There were seven of them on the vessel, which was the most primitive she had travelled in, 'which is saying a good deal. About 25 feet long, she had a tiny open hatch, under which there was just length of space, precariously, to sling a hammock. Her bottom was filled with great ragged stones for ballast.' To the crew and villagers they met, Dolly was 'a source of dazed wonder [...] a white senora, who travelled without a man, who had an incomprehensible desire to see the dread Indios, who actually gave orders and saw to it that they were obeyed – above all, a creature who took baths. Such was beyond human understanding.' Eventually, they got used to her, except her habit of washing twice a day when possible.

To Dolly, the Orinoco was 'the greyest river I have ever seen, it never looks clear or dark, like other rivers', and she came to find Venezuela more remote and unknowable than Africa. The tribes were more elusive, often being hostile to white people, and the country was dense and almost impenetrable. While Africa was a land of many roads, 'of trails made by the feet of millions of black people who walk and walk', here it was very different. White man had not yet been able to take from the Venezuelan Indians their *selva* (forest), said Dolly, 'for the sound reason he cannot travel in it. He can only travel on the rivers.' One could find handsome towns and buildings, wealth and culture 'and all the luxuries and refinements of civilisation side by side with untouched primitivism and unexplored jungle. I know of no other continent in the world where the contrast is so sudden and so marked.'

Over thirty indigenous tribes had been classified in Venezuela at the time of Dolly's visit, several of whom she would encounter, and some of whom were polygamous, others monogamous. In the north-west, she found the Goahiro, whose women were obliged to smear their faces with black paint when they married, because now that they belonged to a man there was no need to be beautiful. She was told that the women allowed a man to take a defective baby to die in a remote spot because, as well as being a family burden, the child when it grew up would otherwise produce inferior children 'to the detriment of the strength and power of the tribe'.

In the tropical forest she encountered the Panare people in their thatched huts, whose customs included a painful ritual for would-be bridegrooms. Once a young man had chosen his partner, he was stripped by the elders and made to wear for fifteen minutes a netted belt containing large stinging ants whose bite was intolerably painful. If he showed no sign of discomfort, he underwent a second test in which the ants were replaced by wasps. In another tribe, it was the man who was credited with childbirth rather than the mother, who had to return to work almost immediately, while the father was feted for eight days, during which he was not allowed to do heavy work or take long walks for fear that exercise might tire the baby's spirit.

For once, Dolly appreciated the positive impact that missionaries – the early ones – had had upon the aboriginal people in the first days of Spanish colonisation and noted that they had played 'the most useful as well as the most charitable part'. The repeated complaints the missionaries made about the ill treatment of the people by the settlers led to the Crown intervening. As well as being men of God, they were men 'of commerce and law, and sometimes, if needs be, of war'. They worked independently of the civil officials and organised centres of teaching and agriculture; they helped the people to protect their settlements and acted as chroniclers, as well as preachers. It was the missionaries who pushed furthest afield and while the faith they taught 'has largely died in the general upheaval of things [...] the great crumbling ruins of their missions still stand where explorer or trader of the Republican system has scarcely ventured'.

The grey, muddy water of the Upper Orinoco flowed deeply 'between never-ending sage-green banks', through the tantalising and impenetrable 'green mansions' of the forest, thought to conceal secret worlds. Her descriptions of the dense green walls that 'seemed even to cage the beasts that lived in it' evoke a creeping sense of claustrophobia. Dolly found the river itself to be 'empty [...] compared with the rivers of other continents, a terribly lonely river'. There were none of the comings and goings of the big African rivers, 'none of the cheerful beehive villages with their chattering inhabitants agog at the sight of a strange craft and new faces'. She missed the cheery banter and exchange of gossip and sometimes found the loneliness of the river 'almost oppressive', so the smallest suggestion of human life – a fishing boat or hunter's hut – was genuinely refreshing.

By contrast, the grey waters beneath them teemed with life. There were *caribes*, 'horrid little cannibal fish' that waited for anything edible to be dropped overboard, and *tonina*, a freshwater porpoise that 'puffed and grunted [...] like an asthmatic old gentleman'. They caught a glimpse of 'the evil head of a water-snake' and once Dolly saw 'the bizarre, glittering flash of a *temblador*, or electric eel', which was dangerous to meet close up, as it could emit a shock sufficient to paralyse a man.

In the towns of Venezuela, she had found the main sport to be cockfighting, which was taken as seriously as racing lovers did the Derby, but on the river the sport was alligator hunting. One evening, her crew wanted to go on a hunt, so she warned them that if she fell in, they would have to pull her out. They replied that there would be no need, as the beasts would reach her before they did.

Eventually, they reached Careño, near the mouth of the Meta River, one of the Orinoco's biggest tributaries, marking the Colombian frontier, where she received the disappointing news that the Meta was almost entirely devoid of water. It meant her quest for a certain area of the country was impossible, and a change of route was necessary. However, she soon realised that for the moment she did not care very much, or indeed about anything at all, for she was feeling too ill. She had succumbed to a fever unlike anything she had known in Africa, 'a brand entirely new to me, a kind of permanent low fever that did not respond to treatment'. Neither quinine nor aspirin seemed to work, and neither did starvation. Although it weakened her, she somehow got used to it. Everyone on the boat fell prey to it eventually, including the captain, so they took it in turns to do each other's jobs.

A little further on was Ayacucho, the frontier town next to Colombia, which she made her temporary base. It was 'the present capital of the gigantic territory of Amazonas; a territory of nearly 18,000 miles with a population of about one person to the square mile'. From here, she could undertake shorter excursions because she realised that due to unavoidable circumstances she had started up the Orinoco at least two months too late. She should have started when the rains were over and there was plenty of water, which would have been November. It was now nearly the end of the dry season, and ahead lay the rapids of Atures and Maipures, which were virtually impossible to navigate.

Her frustration is palpable:

I was on the fringe of the comparatively unknown, of the great territories of Indians who do not like white men, of the great legendary country of the old Conquistadores [...] on the cruel high-road to and across the Brazilian frontier, down to the Amazon. It was tantalising to think of that great tract of country, that great difficult water-road that lay beyond the *raudales* [rapids], natural fortification against the unknown.

With assistance from the governor, Dolly stayed in Ayacucho for two days in a one-roomed hut, trying to regain her strength, repack and reprovision. Outside, in burning heat 'that flayed like a knife', sandflies made a grey, poisonous cloud, and the endless rasp of a gramophone drifted to her during sleepless nights from the neighbouring hut of a soldier and his 'Lady Friend'. The governor lent her his car so she could drive to Maipures on a road (the only one in the region) that skirted the rapids, and then she could voyage along little tributaries. The boat she had been travelling on for the past six weeks was too deep, so she said goodbye to her trusty crew of *El 3 de Diciembre* and drove to a spot above the rapids where the river was low. They hired a small 'cockleshell' of a boat, barely big enough for her and George and a couple of men as paddlers, plus the minimum of food and luggage, and at night, they slept onshore in hammocks slung between trees.

Although they were not navigating the big rapids, they still had to negotiate smaller ones, where the expertise of the paddlers was paramount. Dolly was fulsome in her praise of these men whose skill 'was a miracle of poise, judgment and timing [...] rhythmically perfect [...] with seemingly a sixth sense for the changing violences of the river'. Indeed, they were the only safety barrier between their passengers and the jaws of the alligators and teeth of the *caribes* in the dark waters beneath them.

Yet, despite all the potential dangers, the only direct threat to her life came from an unexpected quarter. One evening, they had set up their little camp by the tributary and she was busy writing when a small boat appeared. Three men got out and walked towards her. They looked in a terrible state, emaciated and dressed in rags. One of them told her in French, with an educated accent, that he had been in the French Army and they had been wandering in the forest for three years. They were trying to get downriver to Ciudad Bolívar but had no food or money.

Dolly got the impression that they were convicts, possibly from the penal settlement in French Guiana, maybe Devil's Island, of whose brutalities she was aware. 'Their crimes may have been of the vilest,' she wrote, 'but now I could not find it in my heart to refuse them. Poor desperate devils, damned souls.' She gave them food and money, quinine and a fishing line and hooks and they made to move off, but the French speaker came back. He resumed talking, telling her she was a brave woman until she felt a sinister element had crept in. She made it clear she no longer wanted to talk, and he left. Her men told her they were Cayennese, that they were known along the river and lived by thieving.

The following day, as her men were eating lunch, she wandered off to take photographs and sat smoking a cigarette. A sudden sound made her look up and there was the man again, this time lunging at her. 'He was unarmed,' she wrote, 'but his hands were thrust out towards my throat and mouth, with a smothering gesture, and there was no mistaking his murderous intention.' Frantically, she fired her pistol at his thigh. He gave a cry but still lurched towards her:

> For a horrible moment the world seemed blotted out by those clutching, smothering hands but a foot or two from my face [...] But then came the sound of stamping feet and fierce cries and up came my men in a rush, shouting hoarsely, brandishing their spears and machetes.

A terrible fight ensued, yet even though he had tried to kill her, she could not stand by and watch her men, three against one, slaughter him. She had to fire several shots into the ground before they stopped, still angry. Even then, 'To leave him in that state under the sun meant death', so she tore up one of her few garments into strips to bandage him as well as she could and sent her men to fetch bread and water. Leaving the still-unconscious figure in the shade, they made for the boat and pushed off quickly. Dolly did not need to cajole them to paddle more quickly, because they worked 'as if all the devils were after them [...] fearful of the possible vengeance of the wounded man's friends'. In all her travels, it was the closest she had come to death at the hands of another human.

TOWARDS SUNSET

It was time to return to Ayacucho and then Ciudad Bolívar – not that Dolly liked returning the way she had come, for it carried with it 'a certain tameness […] The glamour is gone, and the spirit of adventuring.' Everything in life was worth doing once, she said, 'including the drawbacks and the indiscretions, but it takes discrimination to decide what is worth doing twice'. It was tantalising having come so far only to turn back, leaving so much of the river unknown, but she had no alternative. Time was limited and there was no other way, unless circumstances had permitted her to spend another six months 'or a year, or a lifetime, in the Venezuelan bush'. The drawbacks, too, would remain, the discomfort, the heat, the vermin, and she still had a fever and was very tired: little wonder, as she had travelled around 800 miles in very difficult terrain. That she should have got as far as she did was astonishing, even though she had not found, as many people over the centuries had hoped to do, the source of the Orinoco.

But it was a tortuous journey back to Ciudad Bolívar, 'an eternity of idleness and dreary impatience', and her little crew kept fighting private feuds. When they finally arrived, Dolly found a rumour was circulating that she was lost in the wilds 'and cables were flying between official circles on either side of the Atlantic'. Now she had to get to Port of Spain, Trinidad, for the liner back to England and as she said a grateful goodbye to George and others who had helped her, she was touched that a crowd had gathered at the quay to wave her off.

More river travel lay ahead, although this time it was more comfortable until they approached the coast, when they faced the same discomfort as Sir Walter Raleigh had at the point where the waters of the Orinoco 'hurl themselves into the sea'. The flat-bottomed boat was not equipped for it and she felt as if an invisible hand was 'lifting her sharply up and down, dropping her with a bump on the uneven surface of the water, in a manner extremely painful to the anatomies of those of us who were not too busy being seasick', and forcing them to delay dinner until they reached the calmer waters near Port of Spain. There she was greeted by Dr Herbert Spencer Dickey and his wife, who were American and had spent much time in South America, where he had been a doctor for large corporations.

Such was the concern about her that her reappearance sparked great press interest. It may have amused her to know that it put her on the front page of one British newspaper alongside the major story of the Spanish royal family being driven into exile under a new republican government.[1]

As she waited for the liner, she was thrust back into civilised life, 'the smart hotel, the cool potent drinks in long glasses, the good things to eat, the welcoming hum of English voices, the probing fingers of the Press', who would particularly relish her account of being attacked by the fugitive. It felt good, too, to exchange her favoured jungle outfit of 'Oxford bags' – wide-legged trousers that offered the best protection against mosquitoes – for a glamorous frock.

As they steamed away on the liner and the days grew increasingly cool, 'little by little the great land of savannah, of unending forest and gigantic waterways faded into memory'.

Dolly arrived home in mid-April 1931, pronouncing her trip 'the most thrilling ever', and she was soon back in *The Tatler*, wearing high heels and holding a collection of spears. Explaining her love of travel to a reporter, she said, 'I have what might almost amount to a physical and mental claustrophobia, a craving for open spaces, a dislike of being shut in.' Although she liked parties, it was the reason for her 'not being able to stand them in unlimited numbers'.[2]

Soon she completed the book she had been writing en route. *The Country of the Orinoco* was published in January 1932.[3] The *Sunday Times* said it was 'like lifting a curtain on a lost world'.[4] Another reviewer, encountering her books for the first time and thoroughly impressed, felt he had learned much about Venezuela 'without that awful feeling of having been "educated"'.[5] *The Geographic Journal* was more mealy-mouthed, giving little credit for her physical endeavour, although it did allow some praise of the book:

> Though she did not succeed in seeing all that she had intended Lady Dorothy Mills has nevertheless written an interesting account of the Alto Orinoco and the hinterland of Venezuela [...] She has much, almost too much, to say of the discomforts of the journey, and from her descriptions of the region she visited it seems hardly destined ever to be a popular resort. Much of the book is taken up with details of the customs of the Indian tribes, who live in the forests or country round here; and, though there is little that is actually new it is useful to have recent corroboration of the known facts. Its earlier chapters deal with more familiar places [...] and the descriptions of these are picturesque and interesting. The book is well illustrated and its racy style makes it easy to read.[6]

Dolly was thought to be one of only two women who had explored the Orinoco (possibly, the other was Mrs Dickey, although she was with her husband). She was keen that ordinary women should hear of her experiences to widen their horizons and let them know what their sex was capable of. In July, she gave a talk to St Hilda's Girls' Club in Bethnal Green in London's East End, which had been founded for disadvantaged young women. She showed them masks she had collected and may have told them, as she did another group, that the bush 'teaches a woman housekeeping, management, self-control, and all the domestic virtues, and as you can't ring up the stores at the last moment, one's housekeeping has to be 100% efficient'. A forgotten essential item could mean 'an intolerable existence for a dozen or more people for three months or so'.[7] Given that domestic skills did not form any part of Dolly's upbringing – that was what servants were for – it serves as a reminder of her metamorphosis.

As always, she was keen to see her friends, whom she could appreciate all the more for having been away. Her cocktail party at Ebury Street combined an interesting mix of guests, including Robert Cunninghame Graham, a prominent Scottish politician and adventurer; artist Charles Prescott, whose painting of Dolly the earl had bought; Christine Jope-Slade, a novelist; and Francis Sewell Stokes, a screenwriter and playwright who had written a play with Jope-Slade that would shortly open at St Martin's Theatre. Perhaps Arthur was at the party too; she shared with one reporter the morsel that he always hoped she would return home ill after her travels so that she was unable to go away again, 'but I always seem to come back fitter than ever'.[8]

There was one place Dolly would have loved to travel to, but she was born too soon. If she won a big sweepstake, she said (probably thinking of the newly formed Irish Sweepstake), she would 'contemplate fitting out an expedition to the moon. I know that the scientists are all against me, but I believe that with sufficient money for equipment, it could be done.' Realistically, though, she realised she would 'have to win two or three sweepstakes' before she could realise her ambition.[9]

Meanwhile, she had more earthbound issues to deal with. Her father's health had been deteriorating and a nurse kept on for the foreseeable future. Writing from Auckland in December 1930 to Dora at Wolterton, Gladys told her the terrible winter had not helped but now the weather was warmer, Orford was carried into the garden as much as possible, where she hoped they would be able to sleep in a special shelter when the summer got settled. In July 1931, in the middle of another New Zealand winter, he made a codicil to his will and on 27 September he died, at the age of 77. A funeral service took place in Auckland and his body was brought back for burial in Norfolk. As he lay in a lead-lined coffin in the ship's hold, his daughter Anne, now 11, was taken down to see him and to pray for him every day of the three-week voyage.[10]

At the burial at Wickmere on 7 November, the chief mourners were Gladys, now dowager countess, widowed at 39, Dolly, 42, and Anne. In keeping with tradition, his earl's coronet and robes were placed on the coffin, which was carried on a farm wagon to the church, where he was buried with Louise and his two young children. Arthur's attendance is unlikely, given the earl's disapproval of his son-in-law; ten days later, he left for Ghana to gather material for a future serial.[11]

The nature of the relationship between Dolly and her stepmother at that time cannot be ascertained, and the age difference between the half-sisters, as well as their physical separation, was not a recipe for closeness. While grief can have the effect of uniting unlikely individuals, so different were these women's experiences of and emotional connections to the deceased that bonding seems improbable. The earl's will was unlikely to cause a surge of loving feeling in his eldest daughter with its brief, unnamed reference to her: 'My daughter from my former marriage being otherwise well provided for, I make no provision for her under this my will.'[12] It was not that Dolly had expected anything but she must have hoped for a change of heart, however small. Anne was left £5,000, to be invested for her until she was 21, when she was to receive an annual income from it, with the capital going to her children (if any) after her death. Without the longed-for son and heir, the earldom was extinct: instead, Dora's son Bobby now inherited the two baronetcies, becoming 9th Baron Walpole of Walpole and 7th Baron Walpole of Wolterton.

At least Dolly was now able to benefit from her grandfather Corbin's trust fund, which had been created when her mother married, although in real terms the benefit is uncertain. When he referred to it in his will of June 1918, Corbin said the investment (in a US trust fund) was yielding an annual income of £2,000 (approximately £115,600 today). However, when the earl died, the Great Depression was starting in the USA, setting off a roller coaster of economic peaks and troughs, which is likely to have affected the fund's value.

Even if Dolly saw only a small amount, though, it would surely be a welcome addition to her income from writing and may have helped finance her proposed expedition to the Middle East. As Britain held a mandate in the territories she wished to visit, Dolly had to inform the Colonial Office of her plans.

She intended to start at Port Said in Egypt and travel through the Hejaz, the region that covered most of the west coast (on the Red Sea) of what is now Saudi Arabia. One of its towns was Jeddah, the ancient port for pilgrims visiting Mecca, for which she had secured introductions to its governor and to the British representative and hoped they might offer assistance. From there, she hoped to go to Aden, then 'by way of the Hadhramaut coast, Oman and the Persian Gulf to Kuwait and Basra'.[13]

Dolly told the Colonial Office that she knew it would probably be 'quite impossible' to make any part of the journey from Jeddah to Kuwait over land, so she hoped to find some trading ships to transport her. She had visited Iraq before and was not intending to spend much time there, probably taking the route through it operated by Nairn, a New Zealand-owned motor transport company; after that, on to Damascus, Transjordan and Palestine.

She assured the authorities that she would be guided entirely by advice she received in Jeddah and elsewhere in the Hejaz, and understood the risks if she acted contrary to official advice, which she admitted to having done on a previous journey. On that occasion, Sir Percy Cox, a high-ranking diplomat, told her that her action 'might have caused the dispatch of an expedition for her rescue and that she was not worth it'.[14] In the response copied to her, an official said there were no objections to her:

> ... visiting the towns of Muscat, Bahrain and Kuwait and visas for travel to these places may be granted to her at Aden. I shall be glad, however, if you would warn her against attempting to travel in the hinterland of Oman or Kuwait, and against trying to land on the Trucial coast where Europeans are not allowed to land.[15]

She left England in January 1932.

Unusually, she is strangely silent about this expedition. A journalist reported in February that a friend in Jeddah said Dolly was spending time there, but whether she got any further is not known. Perhaps she was advised against it and, having undertaken not to ignore the official line, went no further. Alternatively, she may have been unable to secure sufficient assistance to make her onward journey viable. The sudden death in mid-January of Arthur's father had the potential to disrupt her plans but there is no indication she broke her journey to return to England.

Her invitation to a diplomatically important event in London in May suggests the possibility of her having travelled beyond Jeddah and perhaps reaching Iraq once more. She was part of a large and glittering reception comprising most of the world's ambassadors and foreign ministers in honour of the Emir Faisal, the king of Iraq under Britain's mandate which, by prior agreement, was to terminate in October. The event may also have marked the imminent creation (in September) of the Kingdom of Saudi Arabia, whose name until then was the Kingdom of Hejaz and Nejd, in which Jeddah was situated. While Dolly's invitation was not surprising, given her reputation and her existing diplomatic connections, any recent travel in those areas under the world's spotlight may have made it more relevant.

While Dolly must have been back by the beginning of May, the length of her absence is hard to determine. How Arthur reacted to his wife's return must be inferred from subsequent events. At the start of the year, the focus had been on his father's death and probably his regret that he was unable to join his mother and half-siblings at the funeral in London, apparently because of an attack of influenza (which assumes he was back from Ghana, unless it was a ruse). Perhaps he was able to attend the internment a few days later in their hometown of Bude.

He then had the pleasure of seeing his latest thriller, *One Man's Secret*, about a secret service agent who escapes from Devil's Island, serialised in the *Daily Mirror*, and had started his next book. However, along the line he had made a major life decision which, at some stage – possibly already – involved another woman.

Exactly when Dolly realised their marriage was in serious trouble is unknown, but it was confirmed when Arthur moved out. She started wearing dark glasses, telling a friend at a July wedding that she was rundown since her last trip abroad and suffering from nerves and they were part of her cure. A week later, however, she was admitted to a London nursing home and an announcement made that she was 'suffering from a very bad nervous breakdown and overwork' and was expected to be there for two weeks before going to the country 'for a complete rest'.[16] It is likely that her breakdown was a reaction to her marital situation, although it may have been connected to her previous injury, or perhaps a combination of those things, her body weakened by her accident and her emotions wounded by her husband.

At the beginning of August, she left the nursing home for Switzerland for a rest cure, staying at beautiful Vevey on Lake Geneva, where her friends Freddy and Algy Sladen lived. She was photographed with them, looking frail but relaxed, and clearly enjoying their company and that of their friends, among them the exiled Károlyis and their son Adam.

In early October, she received a letter from Arthur, saying he was not returning to her; it also contained a bill for a hotel in Birchington-on-Sea, Kent. The implication was clear. Women had only very recently been permitted to divorce solely on the grounds of their husband's adultery, of which evidence was required. In providing the hotel bill, Arthur was facilitating the process and showing that he had no intention of contesting her petition. She instructed an investigator to make enquiries, the first step being to obtain photographic evidence. Meanwhile, Arthur was living at a hotel in Brighton.

Dolly's divorce petition, filed at court on 7 November, relied on his adultery at the Beresford Hotel in Birchington 'on or about 30 September 1932 and 1st and 2nd October 1932' with a woman called Jasmine Webster.[17] The requisite photograph of Arthur confirmed him as respondent.

The parties had to attend court for the petition to be heard, which took place on 7 December, when Dolly gave evidence, as did a Miss Agnes Nolan from the Beresford. Perhaps the best thing that could be said was that Arthur's liaison did not take place in some seedy establishment but at one of the top-quality seaside hotels of the time, thus reducing to some extent the element of sordidness that was so often present in such cases, particularly as their break-up inevitably attracted press interest. Nevertheless, it must have been a humiliating experience for Dolly, who had frequently and publicly extolled the desirability of couples pursuing their own interests within marriage.

She moved out of their flat in Ebury Street and into the Ladies Empire Club in Grosvenor Street, a private members' club for women. The decree nisi was issued on 23 January 1933 (the same day as that granted by the same judge to Lady Furness, recently mistress of the Prince of Wales, on the grounds of the adultery of her second husband). A few days later,

Dolly left Liverpool on the *Britannic* for a cruise to Jamaica and, if she was feeling up to it, might have enjoyed some interesting chat, as her diverse fellow passengers included her contemporary, the Duchess of Northumberland, and her two teenage daughters, and Aldous Huxley, whose dystopian novel *Brave New World* had recently been published.

On her return in April, she took a cottage in the country for three months, possibly at Crowborough in East Sussex, for which she had a fondness. It would be comforting to think she was in touch with her stepmother and sister during this difficult time, although in early July they returned to New Zealand.

That month Dolly was legally permitted to file her application for the divorce to be made absolute, which was granted on 25 July. She moved from the Ladies Empire Club (although she would remain a member for some years) into a terraced house in Chelsea, staying in familiar territory.

Arthur was more than ready for a new start. On 4 December in Sussex, he married Monica Cecil Grant Wilson, a doctor's daughter who was fifteen years his junior. He was not her first husband: in 1925, in Australia, she had married Ralph Wilson, a Sheffield man, who had gone out there a few months earlier and she joined him. He divorced her in 1931.

When Dolly wrote her autobiography she thought she had a future – 'I still have a world full of hopes' – but now, at just 44, the combination of her health and divorce deprived her of her old energy. However, although she was no longer in the limelight, she was far from forgotten. At Easter 1933, a newspaper column about the holy season printed a lengthy excerpt from *Beyond the Bosphorus* about her visit to the Sea of Galilee.[18]

Her travel books, along with those of Rosita Forbes, were said in 1934 to be increasingly popular among younger women seeking adventure. During that year, however, Hitler became Führer of Nazi Germany, and Forbes talked to the press about having interviewed him and Mussolini, which would lead to her fall from favour.

Dolly's bon mots were still quoted. Poignantly, she said her favourite saying, attributed to another writer, was '"Never take a return ticket", not only in the travel sense but in the broader issues of life'. Newspapers still relished her work, especially in the USA, which she visited in 1935 to see relatives, including her divorced cousin in Virginia, Amye Walpole

Davis, the daughter of her late uncle, Clare Walpole and his wife Ann, who was just four years older than Dolly. Perhaps they listened together to the radio play *The Devil's Daughter*, adapted from a story she had written for *American Weekly* about spending a night in an Egyptian temple, with British actress Sheelagh Hayes playing Dolly.

In her adventure features for the American market, she deftly combined imaginary detail and dialogue with real accounts of things she had seen and stories she had been told, expanding them into compelling tales and making those foreign lands seem both seductive and thrillingly dangerous. One piece began:

> Not long ago I had the rare luck to spend a month as the guest of a prince of the El Hamri family in his vast fortress-castle high among the mountains of Morocco. What made my visit more than usually intriguing was that my host, Prince Kasra, was preparing for his marriage, not to a lady of his own degree, but to Natasha of the Sloe Eyes, just an ordinary little dancing girl from the desert plains below; and I scented the nucleus of a drama [...] that would probably develop into tragedy.[19]

A feature she wrote about people adversely affected by magic in Portuguese Guinea thrilled American readers by talking of the growth in England of secret societies devoted to Satanism, which she said had recently led to the suicide of a society woman in a London club. It was even affecting her success in finding a tenant for her house in Burnsall Street. One man had withdrawn, saying he had discovered that a 'notorious master of Black Magic' lived nearby 'and he feared for his wife and young daughter'.[20] Dolly does not give the name of this 'master', but it was almost certainly the occultist Aleister Crowley, dubbed 'the wickedest man in the world'. She did not forget her British readers, joining celebrities from the film and theatre worlds, including the pantomime impresario Tom Arnold, in writing for the popular weekly magazine *Ideas*.

The purpose of letting out her Chelsea house was for it to have an occupant while she was away on a four-month Mediterranean cruise beginning in May 1936, taking her maid Louise with her. Not only was it good for her health but it also satisfied to some extent her continued need for travel without all the usual effort. Before leaving she made a will, and in the absence of children of her own, named her cousin Amye's sons, Walpole and Horace, as her main beneficiaries.

Her first stopping place was Marseilles to where, a couple of weeks earlier, Arthur and his wife had also sailed, at an increasingly unsettled time in Europe. In March, Germany had reoccupied the Rhineland, contrary to the Treaty of Versailles, which posed a dilemma to the European allies, especially Britain and France. Arthur anticipated people's fears in his new thriller, *Artists' Model*, a spy series set in Montparnasse, in which the central figure was a beautiful young Czechoslovakian woman and the protagonist a German cavalry officer with a duelling scar on his cheek, who assured her British friends that 'Germany desires war with no nation. But Germany means to be fully prepared for any possible emergency.'[21]

Dolly's last port of call was Corsica, from where she returned in September and went straight to the country, not returning to London until November. Later that month, she announced she was cancelling all engagements to go into a nursing home for an operation.

Still her work was out there. During 1937, her short story 'Romance in the Desert' was included in a book called *Fifty Enthralling Stories from the Mysterious East*, 'by authors whose names are household words', went the blurb. Indeed, they were – they included Somerset Maugham, Joseph Conrad and Sir Walter Scott. Perhaps this rallied her, for in December she sailed from Southampton to Vancouver and stopped off at familiar places including San Francisco and Los Angeles.

A week after her return in March 1938, Hitler annexed Austria and began to draw up plans for Czechoslovakia. Britain's Prime Minister Neville Chamberlain had meetings with him in Germany, the last of which resulted in the notorious Munich Agreement, signed on 30 September by Germany, Britain, Italy and France. Hailed by some as the great hope for peace, it was a divisive document and saw Alfred Duff Cooper, First Lord of the Admiralty, resign in protest.

While Dolly's view of appeasement is unknown, she would not have forgotten what happened to Freddy Sladen when she was in Mussolini's Italy. Certainly, her stepmother was in favour of it. Days after the Munich Agreement was signed, Gladys and Anne, now 19, were returning from a year in New Zealand and en route to New York, on the final leg of their journey, the dowager countess gave an interview in which she was quoted as saying, 'Chamberlain did just right'. The time had come when he 'had to put his pride in his pocket and go meet Hitler face to face'. She was critical of the 'young men [who] wouldn't have done that', referring to Duff Cooper and Anthony Eden, the former Foreign Secretary. Although she had regarded Hitler's word as 'a tricky thing', she was willing to accept Chamberlain's statement that Hitler really wanted peace, 'At any rate we have to believe it because it's all we have to bank on'.[22] A year later, it was a different story.

Anne's marriage in November 1939 gave her and her half-sister something in common: it took place in wartime and Anne's husband was an army officer, Major Joseph Eric Palmer of the Duke of Lancaster's Own Yeomanry. He was fifteen years older than Anne, who, unsurprisingly, had not followed their father's instruction to marry Bobby Walpole – he had married Nancy Jones (at whose wedding Anne and Eric had met) and had an infant son, Robin, to whom Eric was godfather.[23]

Dolly's absence from the Norfolk wedding probably went largely unquestioned, not least because some guests would have been unaware of her relationship with Anne. When the engagement was announced, some newspapers implied that Anne was the late earl's only daughter. That the reception was held at Wolterton would have been too much for Dolly to bear. When her nephews were born, John in 1943 and Anthony in 1945, she would not have expected to be made godmother.

The war restricted travel. Although Dolly still had her Chelsea house, London was not a safe place to live and suffered terribly from the bombings of 1940 and 1941. By 1942, aged 53, she was living in a very comfortable forty-bedroom hotel, Waldronhyrst in South Croydon, which, despite being in a suburban area not known for its beauty, felt as though it was in the countryside, with 6 acres of well-tended gardens producing fruit and vegetables for the restaurant, tennis courts and croquet and putting lawns. Residents like Dolly combined with short-term guests, thus avoiding 'that stagnant atmosphere which is germinated by the same old faces in the same old chairs',[24] as one impressed hotel critic put it. As far as wartime safety was concerned, the hotel was nowhere near a target and in the unlikely event of a blitz in the area, it had its own shelter. Another attraction was that the Waldronhyrst was only half an hour from London's Victoria Station.

In August 1942, eight months after the USA joined the war, Dolly's cousin Amye's son, Horace Davis, now 21, enlisted in the US Army and served in England; his elder brother, Walpole, was in the US Marines and saw action in the Pacific. It is highly likely Dolly made contact with Horace wherever he was based; she would have been the ideal person for pointing him to the best places to enjoy himself in wartime London when he had leave. Later, he would work for the CIA, with foreign assignments that included Baghdad and Latin America, and back home in Charlottesville he founded a travel agency, perhaps using his legacy from Dolly to help realise the dreams of people inspired by explorers like her.

Dolly once said, 'One does not often recognise adventure when it is with one.' Those expeditions she had organised, her tales of which informed and entertained the world, would not happen again, but her work remained relevant. In 1944, during anti-British uprisings by Jews in Palestine, her account from *Beyond the Bosphorus* of Balfour's visit in 1925 was quoted in the US press.[25] At an exhibition for 'Negro History Week' in Mount Vernon, New York, in February 1947, a display of books contained several about Liberia. Dolly's book rubbed covers with Graham Greene's *Journey Without Maps*, an account of his four-week journey (his first outside Europe) through Liberia in 1935, undertaken with a female cousin. Had he read *Through Liberia* first (there is no indication he did), he would have amassed some useful tips.

By 1951, Dolly had moved to the south coast once so favoured by her and Arthur and was living at Steyning Mansions Hotel on Brighton's seafront. Now she retained a nurse, who accompanied her to Portugal for three months to escape the British winter that Dolly had always hated. After her return in 1952, she visited her sister at Rosemoor, an occasion remembered by Anne's youngest son Anthony, although who or what prompted the visit is not known.

They discussed their respective relationships with their father, with Dolly describing hers as tempestuous. There was a time, she told Anne, when she and he were not speaking and as they sat at opposite ends of the long dining table, she communicated by sending him notes via the butler. If she imagined that her sister's upbringing was sweetness and light, however, she would discover differently. Such was the age difference between Anne and the earl that she felt she never really knew him – she was only 11 when he died – and her mother was very strict with her only child. Aspects of Dolly's Victorian upbringing had lingered on in the supposedly modern 1920s, with Anne spending much time in the nursery at Wolterton with a nanny and governess and seeing her parents for just an hour each day. At least Dolly had been able to do things with her parents as she grew older, particularly encouraged by her American mother.

Dolly visited Rosemoor again in 1955, suggesting that the sisters had found common ground, not least perhaps, the knowledge that girls were 'not of interest' to the earl.[26]

In the meantime, on 18 February, Arthur died suddenly in a Hampshire nursing home aged 68, after a brief illness.

The impact of her expeditions remained, with one journalist recalling her account of visiting the Yezidi tribe twenty-seven years earlier and glimpsing the sacred figure of their peacock god. Her return from Guinea was recalled in a column featuring twenty-five-year anniversaries.[27]

On the family side, in the USA, her connection with D.C. Corbin was not forgotten. In 1945, his mansion was bought by the city of Spokane, with an arrangement that allowed his second wife, Anna, to live in it until her death in 1950. By 1956, it had become the Spokane Art Center, celebrated in a lengthy newspaper story that not only paid tribute to Corbin but also talked of the visit of his granddaughter Dorothy when she was 11.[28]

Restless spirit that she was, Dolly did not stay long at Steyning Mansions, although she hardly moved far. Remaining in Brighton, by 1956 she had settled into a hotel converted from a beautiful old house at No. 66 Montpelier Road, away from the seafront.

On 4 December 1959 she died there, at the age of 70. Suffering bronchitis and emphysema – she always enjoyed her cigarettes – she had a stroke. At her request, she was cremated and her ashes scattered over the North Sea off the coast of Weybourne in her beloved Norfolk. Poignantly, she had also made provision for her death should it happen in any other part of the world – cremation, followed by the internment of her ashes 'in the nearest Christian burial ground irrespective of creed'.[29]

For someone who had been so frequently in the public eye, the announcement of her death was muted. Perhaps that was unsurprising, because she had been off the radar for two decades and had no husband or children or even close siblings to remind the world of the extraordinary woman who had left it. On 8 December, *The Times* gave her death a six-line mention in its obituary section rather than in the deaths column, although its brevity hardly constituted a biographical note. Only a couple of local newspapers carried an announcement and even the Norfolk papers were silent.

Yet she knew a lot of people. She had made new friends during her travels and had long-standing friends in the UK, all of whom would have lamented her passing once they heard of it, but of those who were present as she was scattered over the sea there is no known record. It is doubtful her sister or stepmother attended. Her nephew Anthony, by then 14, was not present and as an adult did not remember his mother mentioning it.

A longer tribute in *The Times* on 16 December by an anonymous correspondent went some way to redressing the balance. Headed 'Traveller and Writer', it began poignantly:

The passing of Lady Dorothy Mills has gone by almost unnoticed. Thirty or more years ago she was a figure of considerable note, being an explorer, traveller and author of numerous books.

Readers who had not known her would learn that she was 'a brilliant conversationalist on a wide range of subjects [...] but her favourite topic was Walpoliana', in which she recounted anecdotes about her family 'with dry humour and somewhat merciless wit', although she was always proud of her ancestry. It was tragic, said the writer, 'that a life which had experienced such intrepid adventure and was capable of radiating so much gaiety and humour, should end in self-chosen solitude and comparative loneliness'. The piece maintained that she had retired into seclusion after her accident, preferring 'with her proud and independent spirit, to be forgotten and alone, rather than to be a burden to her relatives and friends'. However, she had not necessarily 'chosen' solitude, and no mention was made of her divorce or the part it surely played in determining her later years.

She had ensured that those relatives and friends she cared about would derive some joy after she had gone. The value of her net estate, after the payment of nearly a third in death duties, was approximately £38,880 (around £1 million). Legacies to friends and employees totalled over £11,000, of which the largest single gift was £8,000 to a woman friend in Kenya. Of her residuary estate, Walpole and Horace Davis each received half and thereby benefited the most financially. However, her best-known legacy was the sum of £1,000 (around £27,600) to the Royal Geographical Society. Dolly had said before her accident:

> When my nomadic habits have left me, my hobby will be with scientific and travel books. And if by then I have made enough money, I hope to be able to help the younger people in their task of discovering and opening up this planet – and possibly, by then, other planets too.[30]

That was what her gift would do, albeit on Earth rather than in space. Dolly specified it was to be used within ten years of her death 'as a grant to any young woman member for some explorative enterprise by land or air and who in the opinion of the Society is most deserving of it'.

In 1965, the recipient was announced as Milada Kalab of Durham University, who the previous year, aged 40, had been appointed the first female lecturer in its Department of Anthropology, which would have delighted Dolly. Born in Prague, she was 14 when the Nazis invaded and after the war, her studies were interrupted by the invasion of the Soviet Union. Her academic prowess and determination led her eventually to England. She used the award for research and fieldwork in Cambodia, which led to her paper *Study of a Cambodian Village*, which was published in 1968 and would result in further important work. In a gratifying continuation of the spirit of her benefactor's legacy and taking a subject that was close to Dolly's heart, on her death in 2016, Milada left much of her estate to Survival International, which helps indigenous people around the world.[31]

On 18 September 2019, sixty years after Dolly's death, her sister died in New Zealand, aged 99. In contrast to the sparse coverage given to the celebrated explorer and writer, Lady Anne Berry (as she was upon remarriage after widowhood) had a full-page obituary in *The Times* which talked of her becoming a noted horticulturist and developing Rosemoor as a significant garden, which she donated to the Royal Horticultural Society in 1988. Dolly was mentioned fleetingly, erroneously referred to as Anne's stepsister. In an interview recorded at Rosemoor in 1976, Anne talked a little of Dolly, including their father's disapproval of Arthur. 'She had a very adventurous life,' she summarised for the interviewer, who knew nothing of her, 'and went where no other white woman had been before.'[32] Of her sister's many achievements, Anne spoke admiringly, almost wistfully, as if regretting she had scarcely known her.

However, the last word must go to Dolly:

As one lives the life of a country and gains the confidence of its people, one learns the loves and hates, the joys and sorrows, the problems that make and mar the lives of all humanity, whether white or black or brown, civilised or primitive, and that are, in the essentials, so very much the same the world over.[33]

EPILOGUE

What makes Lady Dorothy Mills different from other, often better-known women travellers? This book has not attempted to offer a comparison with them, scattered as they are across the centuries, all with different motives and resources and pursuing diverse itineraries. Each was unique, although they all shared courage and determination, as well as the natural advantage their sex gave them. In looking back at these women through the ages, the late Jan Morris, who wrote of having 'the peculiar experience of travelling both as a man and as a woman', concluded that the female traveller has had it easier than the male. 'Women generally offered no threat to anyone,' she wrote. 'Women are still more likely to be helped, women retain their own immemorial methods of persuasion, and most importantly women are more likely to fall among friends, allies or colleagues wherever they go – to this day, the human sorority is stronger by far than the fraternity.'[1]

Dolly herself acknowledged that her sex was often her protection, as most tribes were so unused to seeing a white woman that she was either treated as a novelty or was so outside their experience and comprehension that she barely existed for them. On the occasions when she felt that being female was potentially tricky, the risk of sexual harassment was far more likely to come from other quarters.

One hazard women travellers faced was *because* of their sex – male prejudice, direct and indirect, about which Dolly had much to say. And although she proved herself to be equal to, even better than men in her endeavours and achievements, there was one area in which she would always come second, for the key to succession was being male. In the aristocratic family into which she was born, a son was everything.

Dolly's story distinguishes her in many ways. She travelled during a volatile period, when the world was emerging from the chaos of the Great War and reshaping itself politically and culturally. A century later, it is sobering to see that parts of the Middle East into which she ventured remain deeply troubled, and some other places and peoples she wrote about have, at various times, continued to be the subject of distressing news.

Before Dolly, few women had chosen to journey through Africa, much of which was still little known. That she elected to go there, and everywhere, alone, without a husband, friend or close companion, was exceptional. While her background and title were undoubtedly an advantage when dealing with officialdom, they were of little assistance once in the bush or jungle. Without those who accompanied her at various stages she could barely have made headway: the local people she hired as guides, porters or interpreters, or government officials whose presence she was obliged to accept, at least for part of the way.

As a woman, she achieved many 'firsts' and afterwards shared her experiences with the world by turning them into compelling prose for her travel books and escapist stories for her novels, while undoubtedly being the only explorer who also wrote journalistic features on a wide range of topics for the emerging modern woman. That she continued to enjoy the buzz of a social life when back in England provided a marked contrast with the other worlds she inhabited.

Life is often described as a journey. As an aristocratic woman with all the attendant privileges, Dolly could have taken the easy route, but the consequences of falling in love with the 'wrong' man saw her take another path. In doing so, she underwent a metamorphosis and discovered much about herself, physically and mentally. Although her travels gave her much-needed escape from the worst aspects of 'civilised' society, her heritage remained important to her, but she was deprived of a material part of that when her father gave her childhood homes to another.

Her romantic life also presented challenges. Some women travellers were unmarried – some never had been, a few were widowed – and those who were married often accompanied their husbands. Dolly's situation was unusual. While her fervent belief was that couples should spend time apart, travelling separately from Arthur for months at a time was a risky strategy which ultimately failed. Even her final journey saw her setting off alone, and her ashes scattered to the wind with no one close to her to see her on her way.

A lack of evidence in the form of personal correspondence between the couple leads to unanswered questions. What was Arthur like as a husband? Would he have preferred them to travel together more often? Did they have an open marriage or, apart from his infidelity cited in the divorce, did he – or Dolly – stray? Her writing shows that she was not blind to the physical charms of men, but her autobiography avoids the closely personal. Such absence of detail avoids exposing further than necessary those who had disappointed her, as well as removing the need for overt self-analysis.

Merely to quote from or give summaries of her travel books is to do her an injustice, but until they are republished, it must suffice. There may be phrases she used and opinions she expressed that will jar or offend in today's world, but context is everything. Even in her own time, on at least one occasion Dolly attracted criticism in the press for her manner in writing articles about other races as an upper-class white person – or perhaps as an upper-class white *woman*, because in the 1920s and 1930s most press editors were men.

Notwithstanding her distaste for the public's desire for sensationalism, in trying to give editors what they wanted (and it is impossible to know how much they changed her final copy), her attitude towards those who worked for her could sound patronising. Referring to a specific article of hers, one periodical (whose readership was mainly working-class men) accused her of 'aristocratic gaucherie', although it gave her the benefit of the doubt by saying it was probable that, in trying to provide an article 'with a kick', she had 'distorted the truth'.[2]

Yet, she was always willing to confront her preconceptions, and to have her mind changed. Her sense of humanity is clear. In an age of empire, she referred warmly to her 'black brothers', towards whom she considered that the French had a different and better attitude than the British, giving them 'much more latitude and social rights than we do',[3] as well as being more positive towards those of mixed race. Above all, she understood those aspects of human nature that mankind shared, focusing on what people everywhere have in common rather than what divides them.

Dolly could be critical of her fellow countrymen abroad. At a time when foreign travel was becoming more accessible, she was acutely aware of the poor reputation of the British tourist, whereby a usually modest, pleasant person, once abroad assumes 'the attitude, half-swaggering, half-suspicious, of buccaneers who suspect that someone has designs on their buried treasure!'[4] Once, she watched a large Englishman stride through the hall of a hotel and hector the English-speaking concierge 'in defiantly execrable French', which caused her French companion to remark that the Englishman only travels 'to convince himself that his is the only country in the world'. She also regretted the tendency of upper-class youths to drop their usual sartorial standards when abroad and stare rudely and unreservedly at the locals.

The quality of writing in her travel books distinguishes her and was recognised in her time, not least the humour. Given the panorama of places and peoples she described, it is impossible to select a representative sample, other than the extracts quoted in this book. However, within *A Different Drummer* is her account of a mirage (reproduced here in the appendix), and experienced when travelling with a group of Arabs. At first, it seems as if they all see it, but it quickly becomes apparent that it is only Dolly, whose eyes are 'given vision to see between the twin portals of reality and unreality'. It is a haunting piece that raises many questions. As *A Different Drummer* was written after her father disposed of the Walpole properties, perhaps she retrospectively made a real mirage into a device for describing the emotional effects of becoming an 'outlier'. She had also suffered her accident by the time she wrote the book. Perhaps the mirage, with its shadowy presences, voices and regrets, describes (again, retrospectively) a near-death experience. Whatever her intent, the piece illustrates her belief that not everything can be explained by what mankind thinks it knows.

Dolly may have travelled alone but she was never lonely. 'They are queer things, the friendships of the road,' she wrote:

… oddly enduring, for they are born of solid materials, bred among the stark realities of hunger and thirst, suffering, privation, and sometimes danger. In civilisation one makes friends almost automatically, from force of mutual habit and environment, because one instinctively flocks with birds of the same feather as oneself; but the very mutual incongruity of one's wayside friendships means that they are based on some very definite esteem or sympathy. They are little landmarks of one's nomadic life, living atoms, who help to make up the mosaic entity of the country that one travels.[5]

Only Dolly could have composed such an elegant paean to friends made along the road she was impelled to follow.

APPENDIX

Here is an extract from the chapter 'Written in Sand', from *A Different Drummer* by Lady Dorothy Mills:

'*L'Serab! Le Mirage!*'

The cry dragged me from a state of dazed stupefaction induced by the rhythmic laboured rumbling of the great Renault 'my rover,' under the merciless brilliance of sunshine on brassily-gold shifting sand-dunes, as we approached the Souf. We made up an incongruous over-loaded freight of hot humanity, I and a score of plump, prosperous Arabs, piled up all anyhow, piloted by a French chauffeur who had shared with distinction in a famous trans-Saharan 'Mission' a few years previously, before drink and too much love of living had reduced him to itinerant driver of an Arab firm of cars.

We had started two hours before sunrise, under the stars and a biting little wind that pinched like sharp fingers. Now, the noon was hot as the dawn had been cold; we were all of us thirsty, perspiring, cramped, and a trifle obstreperous, and we had all of us sung, and quarrelled, and yelled a good deal. Some of us had removed our shoes and hung stockinged feet over the side to air, and some of us had not. All round rolled the dunes like a mighty sea, and in the quivering air the leprous patches of greyish scrub seemed to flicker like a pale army of ghosts.

'*Le mirage ...*'

To the left the dunes had ceded to a long expanse of yellowy-grey stony hammada, that stretched as far as eye could see. That much was real anyway, the hammada, and, in the queer way that reality some-times has, reality touched hands with the unreal. Against the farther edge, lay, one would have sworn, a calm sea of rippleless blue water, but a shade paler than the cobalt sky above. It was a long slim bay, studded with small islets, that we looked at, with a low rocky promontory behind, silhouetted against the sky as if cut out of cardboard. At the end of the bay lay a low whitish-pink house, half palace, half bungalow, with two tapering towers, from which a succession of terraces ran down to the bay. Coming in slowly, very slowly, was a ship under half-sail, a ship not of to-day. She seemed a kind of trireme or galleon such as one sees in old pictures that were manned by straining, sweating black slaves, loaded with precious merchandise of gold and sweet spices, and curiously fashioned things of leather, and white-skinned maidens to make pastime for splendid tyrants in the palaces of the north. As one watched, the galleon made her slow way through the calm waters to the lowest terrace that lay before the house.

'Mirage ... a refraction of the sun's rays etc.' So say, in varying phrases, a score of learned tomes and books of reference that know a great deal, know, in fact, all that lies in the brains of the learned men who compiled them. But they do not know everything, neither the learned books, nor those learned men. This was no refraction of the sun's rays that we looked on, nor any other common phenom-enon of science. For in a lull of the babel and guttural profanity of our thirsty, sweating human cargo, my eyes were given vision to see between the twin portals of reality and unreality ...

I knew that I stood on the prow of that galleon, looking ahead with the glad eye of hope towards the pleasant house that was never to be my home. I had crossed many thousand miles of ocean to reach it; my eyes had seen many lands and peoples, I had learnt to understand many things, and now my heart was no longer tired or disillusioned, but fresh, and strong, and glad. And I stepped off the galleon on to the lowest terrace, and went up into the house. I was no longer hot or tired, or dressed in the dusty garments of travel, but bathed and cool and arrayed in soft, rich garments of silk, smooth and grateful to the body. I passed slowly from one room to another, and it seemed to me that in some way the house and its contents were familiar to me, and yet I knew it was a house I had never known before, and shall never know, shall never live in. Slowly, happily, I went down the spacious corridors, examining and appraising. In one room were all the books I had never written and shall never write, books that have lain still-born under my heart, to write which my faith were not great enough, nor my pen skilled, books with messages of power and sympathy and under-standing that should have added a pillar to the corners of the earth. And it seemed to me that round me echoed all the songs I have never sung, and that from the recesses of the air came a chorus of voices low yet acclaiming, echoes of all the praises I have never deserved, of the great deeds I have never performed, of the love of which I have not been worthy. There were other voices, too, voices that questioned and voices whose murmurs seemed about to answer all the things in life that had puzzled and perplexed me.

In another room there were all the dreams that I have ever dreamed, dreams that have not come to fruition, and the hopes that the world and my own weakness have killed. I saw the simple hopes of childhood, the dreams of adolescence and the hopes of tri-umphant youth, the hopes that soar and flame, to whom the world is but a stepping stone to Heaven: and the little hopes of tenderness, of joy and human loving-kindness. And these whispered hopes gave me pleasure, though I knew they were not to be fulfilled.

From the walls of one room, surrounded by a soft light that seemed to make for each a golden frame, looked down on me the faces of those whose friendship I have never known, of those whose eyes have met mine in divers corners of the earth's surface, whose predestined passing had taken something from the lives of each of us. There were smiling eyes, and grave eyes, and thoughtful eyes, and in all of them there was a message that I could not, and shall never, understand. There was a face that had haunted me many days and nights, far out in the East, there was a face that looked so close into mine that its glance was a kiss, there was a pair of eyes that when I had seen them last held tears. And I caught the ghost of a laugh that once had made for me a day of sunlight and dreams …

And there came to me the master of the house, and took me by the hand, and led me through the lofty corridors, and down the silver terraces. He was tall, and grave, and handsome, the man I have never known and shall never know, shadow of my soul, man whose footprints have never crossed mine. And as we walked we did not speak aloud, but our hearts said all the things that had lain in them since the world began, that could never be said by human tongues. And he pointed to where the tall galleon lay, her sails unfurled, swelling to a faint breeze. And we looked at each other and smiled, and before us lay all the greatness of the world, all the wonder and splendour of it that we were not to see together.

A long ripple seemed to run through the calm blue waters, and the galleon stood ready to sail …

'*W'Allah*! *Nom d'un chien*! *Sale piste*!' The moment of silence, crystal-cool, was shattered in the burning immensity. There was a burst of polyglot profanity, as the lumbering camion jarred, skidded and gyrated over the loose sharp-boulders, and lean brown hands darted out to secure scattered belongings. Equilibrium restored, a voice broke out into the high falsetto whine of the usually deep-throated Arab when he sings, and a dry, dust-laden throat expectorated dexterously over the side.

The mirage had faded. Only the sun burnt down pitilessly over an unbroken immensity of rolling brassy dunes.

NOTES

PRELUDE

1 London: W. Heinemann, 1916.
2 Lady Dorothy Mills, *A Different Drummer: Chapters in Autobiography* (London: Duckworth, 1930).
3 George Mills (1896–1972) would become known as a writer of boys' adventure stories, particularly *Meredith & Co.* (1933).
4 He was (later Sir) Mortimer Margesson, husband of Arthur's Aunt Isabel, née Hobart Hampden, his late mother's sister.
5 *A Different Drummer.*
6 *Ibid.*
7 Published by Eveleigh Nash.
8 *A Different Drummer.*
9 *Ibid.*
10 *Ibid.*
11 *Ibid.*
12 *Ibid.*

CHAPTER I

1 *A Different Drummer.*
2 Walpole family papers, WLP 17/10/1, Norfolk Record Office (NRO).
3 *New York Times*, 19 April 1893.
4 Houghton Hall eventually passed to Sir Robert's great-grandson, 1st Marquess of Cholmondeley, through his daughter, Mary. Today it is home to the 7th Marquess and his family.
5 Horatio was created 1st Baron Walpole of Wolterton. His great-great-nephew, Horatio, Admiral Lord Nelson, would later stay at Wolterton.
6 *The Reminiscences of Lady Dorothy Nevill* (London: E. Arnold, 1906).
7 *The Times*, 30 November 1888.
8 Evelyn Caston, *Pall Mall Gazette*, 14 December 1888.
9 *Brisbane Courier*, 7 January 1889.
10 *The Times*, 17 June 1890.

11 He argued that, contrary to the court's view, his lack of reply to her letters in which she said he had promised marriage did not meet that criteria.

12 The delay was blamed on the judge's insistence on having a jury.

13 From *Resignation* by Henry Wadsworth Longfellow (1807–82).

14 *A Different Drummer.*

15 *Ibid.*

16 Quoted in Guy Nevill, *Exotic Groves: A Portrait of Lady Dorothy Nevill* (London: Michael Russell, 1984), p. 172.

17 Letter from Dr Jessop, 20 December 1895, in Ralph Nevill, *The Life & Letters of Lady Dorothy Nevill* (London: Methuen, 1920).

18 Original documents can be found at the Norfolk Record Office.

19 Letter from Lady Dorothy Nevill to Lady Airlie, sent from Maid's Head Hotel, Norwich, undated except '17th' but likely around 1895, in Ralph Nevill, *The Life & Letters of Lady Dorothy Nevill*. It is not certain which 'far-off estate' she refers to. Wolterton was nearby and anywhere much outside Norfolk would be too far for a day out.

20 Quoted in Nevill, *Exotic Groves*, p. 118.

21 *A Different Drummer.*

CHAPTER 2

1 Their stormy marriage did not last long and in 1899 Mary would marry a wealthy English brewer.

2 *A Different Drummer.*

3 A traditional wooden sailing boat used in the eastern Mediterranean.

4 Published in *The Gentlewoman*, 27 June 1914.

5 *Ibid.*

6 *New York Times*, 23 July 1899.

7 Codicil dated 15 March 1897 of the will dated 27 June 1895 of Laura Sophia Frances Walpole, Walpole family papers: WLP 15/48 1049 XI, Norfolk Record Office.

8 The belief that it was Orford who had Amye 'committed' was conveyed to the author by his grandson, Anthony Palmer. A court case of 1903 (about a rent charge on land), in which Amye was a defendant, sheds some light on her situation. Reference was made to the judge to whom her 'lunacy proceedings' were attached and the fact that she was 'not found lunatic by inquisition' (under the Lunacy Act 1890). This meant that to have her treated in an asylum, it was necessary to obtain a legal certificate from a judicial authority. To avoid having someone 'tainted' by a certificate of insanity, there were alternatives (e.g., sending them to a private nursing home), but this was not always legal. It has not been ascertained where Amye was detained.

9 *The Washington Post*, 10 August 1902.

10 *The Gentlewoman*, 21 March 1903.

11 *St Louis Globe Democrat*, 10 November 1903.

12 *The Boston Globe*, 19 August 1903.

13 Estate diary of the 5th Earl of Orford, Norfolk Record Office.

14 Lady Dorothy Nevill, *The Reminiscences of Lady Dorothy Nevill* (London: E. Arnold, 1906).

15 Nevill, *The Life & Letters of Lady Dorothy Nevill*, from a letter she wrote from Ascot in September 1907. Wolterton is regarded as one of England's finest Georgian houses.

16 *The San Francisco Call*, 24 June 1906.

17 *A Different Drummer*.

18 *The Sphere*, 21 July 1906.

19 *The Washington Post*, 14 April 1907.

20 *A Different Drummer*.

21 Tragically, at the age of 37 and suffering from ill health, she would die by suicide.

22 *Omaha Daily Bee*, 27 October 1907.

23 Letter from Lady Dorothy Nevill, written from Ascot, September 1907, in Nevill, *The Life & Letters of Lady Dorothy Nevill*.

24 Mentioned in the *Stamford Mercury*, 2 August 1907, and published in *The Girl's Own Paper* shortly afterwards.

CHAPTER 3

1   *The Tatler*, 29 January 1908.

2   'The Secret of Secrets', interview with Lady Dorothy Mills by Jessie E. Dunbar, *Pearsons Magazine*, November 1919.

3   *Evening Despatch*, 28 May 1908.

4   Estate diary of the 5th Earl of Orford, Norfolk Record Office.

5   The album is in the family's possession.

6   *Eastern Daily Press*, 10 August 1910.

7   *The Kansas City Star*, 3 July 1910.

8   *The Gentlewoman*, January 1910.

9   *Marylebone Mercury*, 12 August 1911.

10 *A Different Drummer*.

11 Estate diary of the 5th Earl of Orford, Norfolk Record Office.

12 *The Gentlewoman*, 1 March 1913.

13 *Ibid.*, 6 September 1913.

14 *Modern Man*, 20 September 1913.

15 *The Atlanta Constitution*, 26 October 1913.

16 Estate diary of the 5th Earl of Orford, Norfolk Record Office.

17 *A Different Drummer*.

18 *The Sketch*, 31 January 1917.

19 *The Gentlewoman*, 10 February 1917.

20 *Cambridge Daily News*, 12 March 1917.

21 *A Different Drummer*.

22 A clasp would be added in 1919, indicating his having operated within range of enemy mobile artillery at the relevant time.

23 He was Somerset Walpole, the eldest of whose three children was the novelist Hugh Walpole (1884–1941).

24 In 1916, the earl bought the Barnaline Estate at Loch Awe. Gladys was staying with friends nearby.

25 Estate diary of the 5th Earl of Orford, Norfolk Record Office.

26 *The Sketch*, 18 September 1918.

27 Laura Arksey, on D.C. Corbin: HistoryLink.org, Essay 7960, posted 10 April 2006 (accessed 4 October 2023).

28 The will of Daniel Chase Corbin, 8 June 1918: Corbin probate documents, archives of the Northwest Museum of Arts and Culture, Spokane (Eastern Washington State Historical Society).

29   *A Different Drummer*.

30   *The Gentlewoman*, 17 May 1919.

31   *The Weekly Dispatch* (London), 8 June 1919.

32   *Ibid.*, 31 August 1919.

33   Estate diary of the 5th Earl of Orford, Norfolk Record Office.

CHAPTER 4

1   *A Different Drummer*.

2   *Ibid.*

3   *Ibid.*

4   *Daily Express*, 10 June 1927, quoted in Lucy Bland, *Modern Women on Trial: Sexual Transgression in the Age of the Flapper* (Manchester University Press, 2013), p. 164.

5   *Daily News* (London), 7 June 1920.

6   Reviews quoted in *Westminster Gazette*, 22 May 1920.

7   *A Different Drummer*.

8   Diary of the 5th Earl of Orford, Norfolk Record Office. The artist was Charles Barrow Prescott (1870–1932). £84 is approximately £4,000 today.

9   *Pall Mall Gazette*, 11 May 1920.

10   *Yorkshire Evening Post*, 26 August 1920.

11   *The Sketch*, 18 August 1920.

12   *The Ladies Field*, 12 November 1921.

13   'Does Man or Woman suffer most in Marriage?', *The Ladies Field*, 31 December 1921.

14   *The Ladies Field*, 19 November 1921.

15   *Ibid.*, 26 November 1921.

16   *Ibid.*, 17 December 1921.

17   *Ibid.*, 31 December 1921.

18   *Ibid.*, 3 December 1921.

19   *Ibid.*, 24 December 1921.

20   *The Daily News*, 5 May 1922.

21   *A Different Drummer*.

22   *Daily Chronicle*, 23 March 1922.

23   *The Times*, 24 February 1922.

24   Her married name became McGrath.

25   *The Mail*, 1 March 1922.

26   *The Sketch*, 15 March 1922.

27   *The New York Herald*, 30 April 1922.

28   The explorer Henry Stanley called Africa 'the Dark Continent' in his 1878 travelogue, remarking that it was poorly known, and it passed into common usage. A more negative understanding of the term, which is now considered pejorative, was that it referred to the nature of the inhabitants.

29   Alice Worsley had visited the caves in 1910, although her experience was less extensive than Dolly's and her account in *Travel and Exploration* magazine suggests she travelled with companions as part of an excursion.

30   Popular with film-makers, the town and surroundings were used for desert scenes in the *Star Wars* film series.

31   'My Adventures among Race of Cave Dwellers', interview, *The Sunday Post*, 28 May 1922.

32   *The Washington Times*, 27 August 1922.

33   *The Courier*, 1 January 1923.

34   *A Different Drummer*.

35   1 November 1922. Holland's real name was Clive James Hankinson. He was also a lecturer.

CHAPTER 5

1   *The Tatler*, 10 January 1923.

2   *A Different Drummer*.

3   *Ibid*.

4   *The Road to Timbuktu* (Boston: Small, Maynard & Company, 1924).

5   *Newcastle Chronicle and North Mail*, 7 November 1922.

6   *The Road to Timbuktu*. Except as otherwise stated, all the quotations in this chapter about her travels in Timbuktu are from this book.

7   *Liverpool Echo*, 1 January 1923.

8   *The Golden Land: A Record of Travel in West Africa* (London: Duckworth, 1929).

9   'She's not good. She chased me.'

10   *A Different Drummer*.

11   Quoted in *The Road to Timbuktu*, pp. 116–17.

12   *A Different Drummer*.

13   From the *Fetach*, quoted in *The Road to Timbuktu*, p. 146.

14   From the *Tarik es Soudan*, quoted in *The Road to Timbuktu*, p. 146.

15   *A Different Drummer*.

16   *A Different Drummer*.

CHAPTER 6

1   She may or may not have been the first *white* woman to enter Timbuktu, because an Irish explorer, Ada Boyland, later said she had bumped into Dolly (who she said she had met before) when they were both there. However, any written accounts of Boyland's journey have not been located and her (brief) interviews to the press about Timbuktu bear marked resemblance to aspects of Dolly's already published experiences.

2   *The Times*, 16 April 1923.

3   *The Birmingham Gazette*, 17 April 1923.

4   *Truth*, 18 April 1923.

5   *The Times*, 22 June 1923.

6   *The Southwark & Bermondsey Recorder*, 18 May 1923.

7   Dolly, speaking to the *Evening News* and quoted in *The Clarion*, 1 June 1923.

8   Dolly to *The Daily News*, 17 April 1924.

9   *A Different Drummer*.

10   Reported in the *Dundee Courier*, 7 April 1924, after her return.

11   *A Different Drummer*.

12   *The Road to Timbuktu*.

13   *A Different Drummer*.

14   Quoted in *A Different Drummer*, p. 174.

15   *The Meriden Daily Journal*, 23 January 1924.

16   *A Different Drummer*.

17 Quoted in *A Different Drummer*, p. 177.

18 *A Different Drummer*.

19 *Ibid*.

20 *The Times*, 25 July 1911.

21 *The Golden Land*.

22 *The Sketch*, 9 August 1924.

23 *The Scotsman*, 13 March 1924.

24 *Daily Mirror*, 11 July 1924.

25 Quoted in *The Daily News*, 4 July, 1924.

26 *Illustrated London News*, 14 June 1924.

27 *The Weekly Dispatch* (London), 21 December 1924.

CHAPTER 7

1 *A Different Drummer*.

2 Now Istanbul.

3 *Beyond the Bosphorus* (Boston: Little, Brown & Co., 1926). All quotations in this chapter about Dolly's travels in the region come from this book unless otherwise stated.

4 'The Empire and the Republic' by Murat Metinsoy, in *History Today*, October 2023.

5 *Ibid*.

6 Now Ankara.

7 *A Different Drummer*.

CHAPTER 8

1 *Beyond the Bosphorus*. All quotations in this chapter about Dolly's travels in the Middle East are from this book unless otherwise stated.

2 In more recent times, the existence of a 'Black Book' has been discovered.

3 *Sunday Post*, 10 May 1925.

4 *The Scotsman*, 9 July 1925.

5 Published by Hutchinson. Review quoted in *The Daily News*, 10 July 1925.

6 *The People*, 24 May 1925.

7 *Buffalo Courier*, 12 July 1925.

8 Cherry, Lady Poynter in the *Lincolnshire Echo*, 1 October 1925.

9 *Through Liberia* (London: Duckworth, 1926). All quotations in this chapter regarding her Liberian expedition are from this book unless otherwise stated.

10 Now usually Sanniquellie.

CHAPTER 9

1 *A Different Drummer*.

2 *Through Liberia*. All of Dolly's quotations in this chapter relating to her travels in Liberia are from this book, unless otherwise indicated.

3 Generally, a residence in the country for a holiday.

4 *Westminster Gazette*, 28 April 1926.

5 *Daily Express*, 22 April 1926.

6 Letter, 22 April 1926, The National Archives, FO 458/91.

7 Letter from Dolly, 23 April 1926, The National Archives, FO 458/91.

8 Letter from Dolly, 26 April 1926, The National Archives, FO 458/91.

9   Handwritten letter from Dolly, 26 April 1926, The National Archives, FO 458/91.
10 Letter, 20 May 1926, The National Archives, FO 458/91.
11 From undated and unnamed press clipping, part of The National Archives, FO 458/91, made in response to her interview in the *Daily Express* of 22 April 1926.
12 *Daily Mail*, 24 April 1926.
13 *Evening News*, 13 August 1910.
14 Raymond Leslie Buell, 'The Liberian Paradox', in *The Virginia Quarterly Review*, Vol. 7, No. 2 (April 1931).
15 Peace A. Medie, *Global Norms and Local Action: The Campaigns to End Violence Against Women in West Africa* (OUP, 2020).
16 *The Tatler*, 5 May 1926.
17 *Daily Mirror*, 5 April 1926.
18 *Daily Mail*, 14 April 1926.
19 *Daily News*, 10 March 1926.
20 Winifred and her siblings, particularly Pamela, feature in the author's book *The Voice from the Garden: Pamela Hambro and the Tale of Two Families Before and After the Great War* (London: SilverWood Books, 2012).
21 *The Scotsman*, 20 November 1926.
22 *The Birmingham Post*, 12 November 1926.
23 *The Sketch*, 30 October 1926.
24 *Westminster Gazette*, 23 October 1926.
25 *A Different Drummer*.
26 Letter, 14 December 1926 in family papers.
27 *A Different Drummer*.
28 *Ibid*.

CHAPTER 10

1   *The Courier and Advertiser*, 25 January 1927.
2   *Daily Mirror*, 20 June 1927.
3   *The Birmingham Gazette*, 28 May 1927.
4   *The Sphere*, 18 June 1927.
5   Attributed to a theory propounded by Edward Carpenter in his 1908 book, *The Intermediate Sex*.
6   *Daily Mirror*, 9 August 1928.
7   *The Tatler*, 10 August 1927.
8   *Daily Mirror*, 2 June 1927.
9   *A Different Drummer*.
10 *Ibid*.
11 *Sunday Mail*, 4 December 1927.
12 Quoted in the *Daily Chronicle*, 5 January 1928.
13 D. Killam, 'Fictional Sources for African Studies', *Journal of the Historical Society of Nigeria*, Vol. 3, No. 2 (December 1965), pp. 377–402. The others were the four African novels of (Arthur) Joyce Cary who, like Dolly, knew West Africa.
14 The pilots were Lieutenant Colonel Frederick 'Dan' Minchin and Flying Officer Leslie Hamilton.
15 'Adventures in Black Magic', *The Graphic*, 21 January 1928.
16 *Daily Mirror*, 19 March 1928.

17 *The Graphic*, 21 January 1928.
18 *The Times*, 14 March 1928.
19 *Belfast Telegraph*, 20 July 1928.
20 Letter from Gladys Orford to Dora Walpole, 16 December 1927, The National Archives: WLP 17/12/24.
21 *Ibid*.
22 *A Different Drummer*.
23 *Deseret News*, 1 September 1928.
24 Letter from Dolly, 19 October 1928, Wellcome Collection.
25 *The Golden Land*. All quotations about her travels in Guinea in the rest of this chapter are from here unless otherwise indicated.
26 *Société Commerciale de l'Ouest Africain*.

CHAPTER 11

1   *The Golden Land*. All quotations about her travels in Guinea in this chapter are from here unless otherwise indicated.
2   Quoted in *The Golden Land*, p. 78.
3   *The Sphere*, 16 February 1929.
4   Reverend L.G. Mannering, MA, MC, *Surrey Mirror and County Post*, 26 April 1929.
5   T.C. Bridges & H. Hessell Tiltman (London: Harrap & Co., 1929).
6   *A Different Drummer*.

CHAPTER 12

1   Letter, 9 July 1929, The National Archives: WLP 17/12/27.
2   *A Different Drummer*.
3   Letter, 1 November 1929, The National Archives: WLP 17/12/27.
4   *Gloucester Citizen*, 30 December 1929.
5   *Illustrated London News*, 18 January 1930. The reviewer was Charles Edward Byles.
6   *The Scotsman*, 13 January 1930.
7   *Daily Mirror*, 26 November 1929.
8   *San Francisco Examiner*, 2 November 1930.
9   Royal Geographical Society, via Wiley Digital Archives. Dolly was elected on 3 November 1930.
10 *Evening Sentinel*, 2 December 1930.
11 Lady Dorothy Mills, *The Country of the Orinoco* (London: Hutchinson, 1931). All the subsequent quotations in this chapter about her journey in Venezuela are from this book unless otherwise stated.

CHAPTER 13

1   *Evening Sentinel*, 15 April 1931.
2   *Daily Mirror*, 12 May 1931.
3   Lady Dorothy Mills, *The Country of the Orinoco* (London: Hutchinson, 1931).
4   Quoted in the *Yorkshire Post*, 20 January 1932.
5   *Sheffield Daily Telegraph*, 11 February 1932.
6   *The Geographical Journal*, Vol. 80, No. 1 (July 1932), pp. 81–2.
7   *The Northern Whig and Belfast Post*, 30 July 1931.
8   *Ibid*.

9   *The Lancashire Daily Post*, 1 June 1931.
10 As told to the author by Anthony Palmer, Lady Anne's son.
11 'Black Royalty', *News Chronicle*, 24 April 1933.
12 Earl of Orford's will, 3 November 1928. WLP 17/12/30, Norfolk Record Office.
13 Memo to Colonial Office signed by J.W. Stafford, 16 December 1931, 19/4 Lady Dorothy Mills and Others [7r] (13/34): British Library: India Office Records and Private Papers IOR/R/15/2/592, in *Quarter Digital Library*, accessed September 2022.
14 *Ibid.*
15 Letter from British Residency, Bushire, signed by H.V. Biscoe, 12 February 1932: British Library: India Office Records and Private Papers. The Trucial Coast was the name given by the British Government to a group of tribal confederations to the south of the Persian Gulf whose leaders had signed protective treaties with the UK. They later formed the United Arab Emirates.
16 *The Times*, 16 July 1932.
17 Divorce and Matrimonial Cause files, ref. J 77/3091/5200, The National Archives.
18 *The Blackburn Times*, 15 April 1933.
19 'Like an Arabian Nights Adventure', *San Francisco Examiner*, 23 September 1934.
20 'Seized by a Black Magic Demon', *San Francisco Examiner*, 23 February 1936.
21 Chapter 1 of *Artists' Model*, quoted in *Daily Mirror*, 28 September 1936.
22 *The Salt Lake Tribune*, 5 October 1938.
23 Robin (Robert Horatio Walpole, 1938–2021) would succeed to his father's titles in 1989. In October 2023, the author met his widow, Laurel, Lady Walpole, (his second wife) at Mannington Hall, which remains in the family. Wolterton was sold in 2016.
24 Ashley Courtenay, in *The Sphere*, 2 August 1941.
25 *Star Tribune*, 10 March 1944, the largest newspaper in Minnesota.
26 As suggested by Anthony Palmer to the author.
27 *Montrose Standard*, 22 April 1954.
28 In the 1970s it officially became the Corbin Art Center, which it remains today.
29 Dolly's will dated 26 May 1936.
30 *Daily Mirror*, 29 October 1930.
31 From an information sheet on the late Milada Kalab, posted by Durham University Anthropology Department on Twitter/X, 8 March 2019, for International Women's Day.
32 Recorded interview by Sue Scutton, June 1976, from Bideford & District Community website, accessed 1 March 2024. Lady Anne was still Palmer at that point. Her husband died in 1980, and she married New Zealander Bob Berry in 1990.
33 *A Different Drummer*.

EPILOGUE

1   From the foreword to *Off the Beaten Track: Three Centuries of Women Travellers* by Dea Birkett (London: National Portrait Gallery, 2004). Jan Morris (1926–2020) was born James Morris and transitioned in 1972. She was a historian, traveller and journalist.
2   *John Bull*, 11 August 1928.
3   *The Road to Timbuktu*.
4   *The Daily Chronicle*, 2 May 1928.
5   From 'A Wanderer's Friends' in *A Different Drummer*.

ACKNOWLEDGEMENTS

I should like to give warm thanks to the following people who have helped to make this book possible:

Lady Antonia Fraser; Anthony Palmer, for his enthusiasm, support and time, permissions to quote from Dolly's works and use of photographs; Laurel, Lady Walpole, for showing me Mannington Hall and for photographs and permissions; Peter Sheppard and Keith Day, for their generous hospitality at Wolterton Hall (when it was in their ownership), which they had painstakingly renovated; Angus Sladen, for photographs of Dolly, his grandparents and friends; Dr Jennifer Binczewski, for facilitating from Spokane my research on D.C. Corbin; the staff at the Norfolk Record Office; Claire Hartley at The History Press; the Society of Authors; the Royal Geographical Society; and my husband, as always, for his unfailing support and encouragement.

SOURCES

ARCHIVES

The British Library – magazine features by Lady Dorothy Mills; India Office records concerning her visit to the Middle East, 1932.

The National Archives – Arthur Mills' army record; correspondence concerning the Liberian expedition 1926; Divorce and Matrimonial Causes file.

Norfolk Record Office – Walpole family archives.

Royal Geographical Society – documentation regarding fellowship.

Spokane County Archives – D.C. Corbin papers relating to his will and estate.

Wellcome Collection – correspondence from Lady Dorothy Mills.

OTHER PRIMARY SOURCES

Author's interviews with Anthony Palmer (nephew of Lady Dorothy Mills) and Laurel, Lady Walpole.

Interview with Lady Anne Palmer by Sue Scutton.

Ladies' Empire Club Members' List, 1936 (accessed via London School of Economics Women's Library digital resources).

National Register of Historic Places (US Department of the Interior, National Park Service).

Newspapers, British and American (accessed via searchable online repositories) – publications are named within the main text and in the notes.

Private family letters.

Public documents – certificates of birth, baptism, marriage and death, census returns, electoral rolls, passenger lists, wills and probate documents.

Wolterton Visitors' Book.

BOOKS AND ARTICLES

Birkett, Dea, *Off the Beaten Track: Three Centuries of Women Travellers* (London: National Portrait Gallery, 2004).

Bland, Lucy, *Modern Women on Trial: Sexual Transgression in the Age of the Flapper* (Manchester University Press, 2013).

Bridges, T.C., & H. Hessell Tiltman, *More Heroes of Modern Adventure* (London: George G. Harrap, 1929).

Buell, Raymond Leslie, 'The Liberian Paradox', *The Virginia Quarterly Review*, Vol. 7, No. 2 (April 1931).

Bush, Barbara, 'Britain's Conscience on Africa: White Women, Race and Imperial Politics in Interwar Britain', in *Gender and Imperialism*, edited by Clare Midgely (Manchester University Press, 1998).

Dismore, Jane, *The Voice from the Garden: Pamela Hambro and the Tale of Two Families Before and After the Great War* (London: SilverWood Books, 2012).

East, D., G. Bentley, & P. Sillitoe (Durham University Anthropology Department), 'Making Women Visible', information sheet on Milada Kalab posted on Twitter/X (8 March 2019).

The Geographical Journal, Vol. 80, No. 1 (July 1932), pp. 81–2.

Kershner, Jim, 'Noble Branch: A British Historian Looks into Anglo-American Romance Between Louise Corbin and Lord Walpole', online feature for *The Spokesman Review*, posted 11 March 1996, updated 16 July (last accessed 15 May 2024).

Killam, D., 'Fictional Sources for African Studies', *Journal of the Historical Society of Nigeria*, Vol. 3, No. 2 (December 1965), pp. 377–402.

Leng, Kirsten, 'Permutations of the Third Sex: Sexology, Subjectivity, and Antimaternalist Feminism at the Turn of the Twentieth Century', *Signs*, Vol. 40, No. 1 (Autumn 2014 from JSTOR).

McKenzie, Scott, *A Brief Guide to St Andrew's Church, Wickmere* (2011).

Medie, Peace A., *Global Norms and Local Action: The Campaigns to End Violence Against Women in West Africa* (Oxford University Press, 2020).

Metinsoy, Murat, 'The Empire and the Republic', *History Today*, October 2023.

Mills, Arthur, *Artists' Model* (fiction series), *Daily Mirror*, September 1936 (other works by Mills are referred to in the text).

Mills, Lady Dorothy, *Card Houses* [fiction] (London: Eveleigh Nash, 1916).

— *The Laughter of Fools* [fiction] (London: Duckworth, 1920).

— *The Tent of Blue* [fiction] (London: Duckworth, 1922).

— *The Road* [fiction] (London: Duckworth, 1923).

— *The Arms of the Sun* [fiction] (London: Duckworth, 1924).

— *The Road to Timbuktu* [travel] (Boston: Small, Maynard & Company, 1924).

— *The Dark Gods* [fiction] (London: Duckworth, 1925).

— *Beyond the Bosphorus* [travel] (Boston: Little, Brown & Co., 1926).

— *Phoenix* [fiction] (London: Hutchinson, 1926).

— *Through Liberia* [travel] (London: Duckworth, 1926).

— *Episodes from 'The Road to Timbuktu'* [travel] (London: George Harrap, 1927).

— *Master!* [fiction] (London: Hutchinson, 1927).

— *Jungle* [fiction] (London: Hutchinson, 1928).

— *The Golden Land: A Record of Travel in West Africa* [travel] (London: Duckworth, 1929).

— *A Different Drummer: Chapters in Autobiography* [autobiography] (London: Duckworth, 1930).

— *The Country of the Orinoco* [travel] (London, Hutchinson, 1931).

Nevill, Lady Dorothy, *The Reminiscences of Lady Dorothy Nevill* (London: E. Arnold, 1906).

Nevill, Guy, *Exotic Groves: A Portrait of Lady Dorothy Nevill* (London: Michael Russell, 1984).

Nevill, Ralph, *The Life & Letters of Lady Dorothy Nevill* (London: Methuen, 1920).

Shattuck MD, George C., 'Liberia and the Belgian Congo', a paper read at the meeting of the Society on 21 May 1928, from the *Geographical Journal*, Vol. 73, No. 3 (March 1929).

POEMS

Poems by Lady Dorothy Mills, in the name of Walpole (those referred to in the text and notes and published in newspapers/magazines):

'A Buttercup' (1907).
'A Type' (1913).
'Doubts' (1913).

SHORT STORIES

Short stories by Lady Dorothy Mills (a sample of those published in newspapers/ magazines and referred to in the text and notes). Those pre-dating 1916 are in the name of Walpole:

'A Fable' (1913).
'The Call' (1914).
'Where Anything Might Happen' (1920).

WEBSITES

Arksey, Laura, on D.C. Corbin: HistoryLink.org, Essay 7960 (posted 10 April 2006, accessed 4 October 2023).

Bideford & District Community online archive: Lady Anne Berry interview (accessed 20 December 2023).

Book Riot (James Wallace Harris): 'The Resurrection of Lady Dorothy Mills' (accessed 3 July 2018).

Kershner, Jim, 'Noble Branch', 11 March 1996: *The Spokesman Review* website (accessed 23 March 2023).

Peerage.com (last accessed 22 May 2024).

'Who is George Mills?' (Harry Williams): whoisgeorgemills.com (comprising several parts, last accessed 25 March 2024).

INDEX